NIKOLA TESLA: GUIDED WEAPONS & COMPUTER TECHNOLOGY

Electrical Review (N.Y.), March 29, 1899

East Houston Street laboratory, New York. (Demonstration oscillator described in Interference Transcript at pp. 17-19.)

Tesla Presents Series
Part 3

NIKOLA TESLA: GUIDED WEAPONS & COMPUTER TECHNOLOGY

Leland I. Anderson, Editor

1998

TWENTY FIRST CENTURY BOOKS BRECKENRIDGE, COLORADO

 Published 1998
Printed in the United States of America
03 02 01 00 99 6 5 4 3 2

ISBN 0-9636012-5-3 (hard cover)
ISBN 0-9636012-9-6 (soft cover)

Library of Congress Cataloging-in-Publication Data

Tesla, Nikola, 1856-1943.
Nikola Tesla : guided weapons & computer technology / Leland I. Anderson, editor.
p. cm. -- (Tesla presents series ; pt. 3)
Includes bibliographical references and index.
ISBN 0-9636012-5-3 (hardcover). -- ISBN 0-9636012-9-6 (pbk.)
1. Tesla, Nikola, 1856-1943--Trials, litigation, etc.
2. Fessenden, Reginald Aubrey, 1866-1932--Trials, litigation, etc.
3. Patent suits--United States--History--20th century. 4. Remote control--History--20th century. I. Anderson, Leland I. II. Title.
III. Series.
KF228.T427T47 1998
346.7304'86--dc21 98-22686
CIP

Twenty First Century Books
P.O. Box 2001
Breckenridge, Colorado 80424

Contents

Appendix

Figures

Part 1

Part 2

From the Publisher

Preface

Nikola Tesla was born of Serbian parents at Smiljan, in the Austro-Hungarian border province of Lika, now part of Croatia, at midnight July 9-10, 1856. His father, Milutin, was a Serbian Orthodox priest, and his mother, née Djouka Mandić, was of a family line whose sons were of the clergy and whose daughters were wives of the clergy. The Serbian Orthodox church then used the Julian calendar, and it continues to use this calendar today for days of observance. The American colonies had converted to the Gregorian calendar 132 years before Tesla arrived at New York in 1884. When Tesla crossed this 'date line,' 11 days dropped from his personal calendar. Most institutions observe Tesla's birth date as July 10, which date Tesla held for himself, but if the tolling church bells in Lika could have been heard in America when Tesla was born, the calendar date would have been July 21, 1856.

Establishing himself in the United States, Tesla became a citizen in 1891. He brought to the world great inventive gifts for which he became famous—the induction motor and the multi-phase alternating current power distribution system driving it (1888); the fundamental system of wireless telegraphy, embodying the "Tesla coil" (1893); telemechanics (1898); the Tesla turbine (1913, 1920), which is attracting great revival interest; and, among other leading inventive achievements, VTOL aircraft (1928).

For 50 years following presentation of the principles of wireless telegraphy, now called radio, during his demonstration lecture in St. Louis, Tesla steadfastly asserted his inventive claim. It was not until five months following his death at the age of 86 in 1943 that the U.S. Supreme Court declared the basic radio patent of Marconi invalid, recognizing the prior art of Tesla for the system concept and apparatus, John Stone Stone for the method of selectivity, and Oliver Lodge for variable tuning.

In his lifetime, Tesla was granted over 30 honorary degrees and foundation medals from the world over. The unit of magnetic flux density in the MKS system was named “tesla” on the occasion of the centennial year of his birth. The only other American to share such recognition is Joseph Henry.

This book is based on and includes the depositions transcript of the U.S. Patent Office Interference between Tesla and Reginald Fessenden for a patent now known as the fundamental AND logic gate.* The depositions transcript document has not heretofore been available. A copy was found among a group of papers passed on to his children by George Scherff, Tesla’s secretary and one of the three giving a deposition. The U.S. Patent Office copy was discovered missing from official records decades ago. Another copy is reported to exist in the archives of the Tesla Museum in Belgrade, but it has been unavailable to scholars.

In retranscribing the document (in poor condition), occasional typographical errors were noted. For example, articles “a” and “an” were corrected from “o” and “in”; the number “ten” corrected from “tin”; “than” corrected from “then”; and missing periods and capitalizations entered. There were no significant stenographic errors. As an aid to the reader, explanatory notes are provided for certain items in the text.

—— Σ ——

* In patent law, the term Patent Office Interference "designates a collision between rights claimed or granted; that is, where a person claims a patent for the whole or any integral part of the ground already covered by an existing patent or by a pending application. The term refers to a Patent Office proceeding to determine priority of invention between two or more parties claiming the same subject matter." (*Blacks Law Dictionary*, St. Paul, Minn., West Publishing Co., various eds.) In the proceeding, the contending parties appear before a patent examiner and, based on evidence presented in terms of deposition(s) and exhibit(s), the examiner renders a decision on priority.

Introduction

Nikola Tesla's technological legacy

Until recent years, Nikola Tesla's legacy in the engineering sciences has been largely unknown. Not associating himself with corporate enterprise or academic institutions, he embarked on a course of independent inventorship. Such scientist-inventors, with courage, ambition, and indefatigable spirit, have a monumental challenge in also managing the promotion of their work in the competitive industrial marketplace. These technologists must also be wary of giving away their income sources—patent concepts—in public appearances and lectures announcing their achievements. Tesla delivered nine historic lectures and papers in rapid succession before the turn of the century as shown in the table that follows.

Lectures and Papers Delivered by Nikola Tesla
1888 through 1899

"A New System of Alternate-Current Motors"
May 16, 1888, before a meeting of the American Institute of Electrical Engineers (AIEE) in New York City.

- * -

Trio series demonstration lectures on high frequency alternating currents, 1891-1893

"Experiments with Alternate Current of Very High Frequency and Their Application to Methods of Artificial Illumination"
May 20, 1891, before a meeting of the AIEE in New York City.

"Experiments with Alternate Currents of High Potential and High Frequency"
February 3, 1892, before the Institution of Electrical Engineers, London,
February 4 (with some condensation), by special request at the Royal Institution, London,
February 19, by invitation, before the Société Internationale des Electriciens and the Société Francaise de Physique, Paris.

"On Light and Other High Frequency Phenomena"
February 24, 1893, before the Franklin Institute in Philadelphia,
March 1 (with some variation), before a meeting of the National Electric Light Association in St. Louis (it was in the latter that the principles of radio telegraphic communication were first presented).

- * -

"Mechanical and Electrical Oscillators"
August 25, 1893, before a meeting of the International Electrical Congress at the Columbian Exposition in Chicago,
November 29 (with some variation), before a meeting of the New York Electrical Society.

"On the Streams of Lenard and Roentgen, with Novel Apparatus for Their Production"
April 6, 1897, before the 80th meeting of the New York Academy of Sciences.

"Notes on an Electrical Oscillator"
August 24, 1897, a paper read before the 67th meeting of the British Association for the Advancement of Science, Toronto.

"High Frequency Oscillators for Electro-therapeutic and Other Purposes"
September 13, 1898, a paper read before a meeting of the American Electro-Therapeutic Society in Buffalo.

"Telautomatics"
May 13, 1899, before a meeting of the Commercial Club in Chicago (Tesla was then en route to Colorado Springs to conduct experiments in wireless power transmission and in this lecture he demonstrated the operation of his radio-controlled boat, a "telautomaton," on water).

The demonstration lectures were feats of brilliant technical display. Lecture halls were filled to standing room only. Tesla was presented with thunderous ovations and wreaths in appreciation. He shipped curious arrays and assortments of apparatus—glass bulbs and tubes, and oil-filled condenser and transformer tanks—across country and ocean. In all the reviews of these demonstration lectures, not one mention of demonstration failure.

It is rather remarkable that Murphy's Law† seemed not to apply to him.

But the extreme workload and financial burden of seeking patents on discoveries and technologies presented in the lectures began to take its toll in missed opportunities. Others took advantage.

After the turn of the century, Tesla's lectures became less frequent and his work products were disclosed only after patent protections had been secured. A half-century after his death, his unpublished writings and interview texts are becoming available and appearing in published form. Information of surprising current relevancy is disclosed; his writings are attracting interest among the communities of physical science, engineering, and the general public. Clearly, few instances of such posthumous rediscovery and later embrace of a scientist's work are known.

The setting

In his laboratory on South Fifth Avenue (now West Broadway) in New York City, Tesla began work in the 1890s on the construction of remote-controlled devices, or "telautomatons," as he referred to them. This work evolved from his investigations in 1893 of automatic mechanisms actuated from a distance.†† Tesla performed experiments with tuned devices (secondary-actuated) that would respond by induction to an oscillator primary coil running up and around the cove in the experimental area of the laboratory.

† "As is well known to those versed in the state of the art, Murphy's Law states that *If anything can go wrong it will . . .* Some authorities have held that Murphy's Law was first expounded by H. Cohen when he stated that *If anything can go wrong it will—during the demonstration.* However, Cohen has made it clear that the broader scope of Murphy's general law obviously takes precedence." From, a tongue-in-cheek article by D.L. Klipstein in *IEE* (Denver), Aug. 1967 (Vol. 15, No. 8), "The Contribution of Edsel Murphy to the Understanding of the Behavior of Inanimate Objects."

†† Tesla, in Part VI—"The Art of Telautomatics," from a six-part series entitled "My Inventions" appearing in the *Electrical Experimenter*, Feb.-June and Oct. 1919 issues. This series has been republished as *Moja Pronalasci—My Inventions*, Školska Knjiga, Zagreb, 1977, on the occasion of Tesla's 120th anniversary, in English side by side with a Serbo-Croatian translation by Tomo Bosanac and Vanja Aljinović, Branimira Valić, ed., and as *My Inventions*, Hart Bros., Williston, Vt., 1982, Ben Johnston, ed.

Tragedy struck March 13, 1895, when the laboratory was burned to the ground. Charles Dana, editor, the New York *Sun*, remarked in its columns,

> The destruction of Nikola Tesla's workshop, with its wonderful contents, is something more than a private calamity. It is a misfortune to the whole world. It is not in any degree an exaggeration to say that the men living at this time who are more important to the human race than this young gentleman can be counted on the fingers of one hand; perhaps on the thumb of one hand.

Driven by the enormous possibilities his investigations suggested, Tesla lost no time in establishing another laboratory on East Houston Street. He was now able to have tuned devices respond *conjointly* by induction from the coil around the experimental area. A very interesting prospect indeed! The question remained, could this condition also be achieved by radio (then wireless telegraphy)?

The 'statics' from various sources in the city—DC motors, contactors, elevated railways, and other power sources—made it most difficult to attempt wireless duplication of the induction experiments. In June of 1899, he established an experimental station at Colorado Springs for wireless investigations. The site, on the eastern plains of the city, was ideal. However, when beginning investigations, he found that static interference was present their too.[†] It should be noted that many researchers and amateurs were then experimenting with wireless telegraphy. In this connection, an interesting question has been raised about the alleged Marconi transmissions across the Atlantic, March 12 and 13, 1901. Robert M. Marriott, first president of the Institute of Radio Engineers remarked (in a deposition taken March 26, 1925):

> Insofar as Marconi's claim of getting the letter *S* [three successive dots] across the Atlantic, which, I believe was the first incident claimed, I have always been more or less lost as to what to think about it, because (at) very nearly the same time

[†] Anderson, L., ed., *Nikola Tesla On His Work With Alternating Currents and Their Application to Wireless Telegraphy, Telephony, and Transmission of Power* (Denver, Sun Publishing, 1992), p. 167.

> I made daily experiments with the coherer receiver on the Atlantic coast, and got such multitudes of dot letters, the letter *S* by the hundreds, two-dot letters in greater numbers, and one-dot letters in still greater numbers. . . .

Consequently, Tesla vigorously addressed the problem as an extension of the experiments performed in his New York City laboratories. In preparation for work to come, on June 27, 1899 he sketched out a set of circuits for the conjoint transmission and reception of signals at different frequencies.[†] So important did he consider the individualization concept and technique that following his return to New York City in January 1900 he instructed Scherff, his secretary and useful man, as a first order of business, to prepare patent applications from the notes kept in Colorado Springs.

The patent application was submitted, but during the period awaiting review, Tesla received notification from the U.S. Patent Office that another patent had been received from Reginald Fessenden on a similar concept and that the office was directing an Interference investigation to determine priority. Depositions were taken in 1902, and the following section is a transcript of those depositions.

In addition to resolving the issue of Tesla's patent application having priority over that of Fessenden, the depositions are of interest today because they disclose not only unavailable information about the development of the invention but as well heretofore unavailable information about experiments performed at Tesla's laboratory in New York City and his experimental station at Colorado Springs.

A discussion and analysis of these disclosures is presented in Part 2.

[†] Marinčić, Aleksandar, ed., *Nikola Tesla - Colorado Springs Notes, 1899-1900* (Belgrade, Yugoslavia: Nolit, 1978).

Part 1

In the United States Patent Office

Nikola Tesla

vs.

Reginald A. Fessenden

Interference No. 21,701

Systems of Signaling

Boyle & Anderson Archives

INDEX

Exhibits Where Offered in Evidence

In the United States Patent Office

Nikola Tesla vs. Reginald A. Fessenden	Interference No. 21,701. Systems of Signaling.

Application of Nikola Tesla, Serial No. 23,847, filed July 16, 1900, and application of Reginald A. Fessenden, Serial No. 18,878, filed June 2, 1900.

Preliminary Statement of Nikola Tesla

Nikola Tesla, of New York, in the County of New York and State of New York, being duly sworn, deposes and says that he is a party to the interference declared by the Commissioner of Patents April 1, 1902, between his application for letters patent filed July 16, 1900, Serial No. 23, 847, for System of Signaling, and an application of Reginald Fessenden; that he is unable at the present time to fix the precise date on which he conceived the invention set forth in the declaration of interference prior to the 1st day of May, 1898; that he is unable at the present time to fix the precise date on which he first made a drawing of the invention, for the reason that he is constantly making sketches of the inventions on which he is working, but that he knows he made drawings of the said invention prior to the 1st of May, 1899; that he disclosed the invention to others prior to the 15th day of June 1899; that he reduced the invention to practice at least as early as the 1st of May, 1898; that he has made no model as distinguished from a practical operative machine; and that the invention has not been used commercially up to the present time.

Nikola Tesla

New York, August 5, 1902

Before the Commissioner of Patents in the matter of an interference between the application of Nikola Tesla, filed July 16, 1900, Serial Number 23,847,[a] for, Systems of Signaling, and an application of Reginald A. Fessenden, filed June 2, 1900, Serial Number 18,878, for Improvement in the Transmission and Receipt of Signals, Interference No. 21,701.

Depositions of witnesses examined on behalf of Nikola Tesla pursuant to agreement at the office of Kerr, Page & Cooper, 120 Broadway, New York City, on August 5, 1902.

Present — Parker W. Page, Esq., on behalf of Tesla; and Bayard H. Christy, Esq., on behalf of Fessenden.

Met pursuant to agreement, formal notice being waived.

The above interference being one of two companion patents, the second numbered 21,702, between applications of the same parties, that of Tesla's being application No. 64,522,[b] filed June 14, 1901 (a division of the application above named), and that of Fessenden being application No. 18,878, filed June 2, 1900, the said applications being for methods, it is stipulated and agreed that the testimony taken in this case may be entitled and used in both cases.

[a] Refer to Appendix D-2 for resulting patent No. 725,605.

[b] Refer to Appendix D-1 for resulting patent No. 723,188.

NIKOLA TESLA, a witness called on his own behalf, having been first duly sworn and says, in answer to interrogatories proposed by counsel, as follows:

1 Q. Please state your name, age, occupation, and residence.

A. Nikola Tesla; Waldorf-Astoria, New York City; place of occupation, Wardenclyffe, Long Island; age, 44;* occupation, electrician.

2 Q. You are one of the contestants of this interference?

A. Yes.

3 Q. Have you read the applications of Reginald A. Fessenden involved in this interference, and do you understand the same?

A. I have, and understand the same.

4 Q. Have you read the stated issues of the interference as contained in the Patent Office notices, and do you understand the subject matter which is in dispute?

A. I have read and understand them.

5 Q. Will you give a general history of your work in connection with methods and appliances pertaining to the general subject matters of this interference?

A. As nearly as I recollect, I entered the study of electrical oscillations and apparata for producing the same some time in 1889. I saw great possibilities and concentrated my efforts on producing efficient apparata and devising methods for various purposes. I think I gave the first demonstration with tuned circuits before the American Institute of Electrical Engineers in 1891. This work I followed up, and subsequently I presented novel and advanced results before scientific societies abroad in

* Stenographic error; should be 46.

1892. My experiments elicited unusual interest, particularly those which I performed with a novel form of oscillatory apparatus, which scientific men have since generously identified with my name. Some time in 1892, on my return from abroad, I constructed and improved apparatus for the purpose of carrying on my practical research of tuned circuits, and I believe in 1893 I lectured before the Franklin Institute and the National Electric Light Association and recorded results still further advanced along the same lines. In these lectures I gave the first outline of a system of transmitting intelligence. Subsequently, I turned my efforts to developing these ideas and results along practical lines. These efforts resulted in numerous improvements which I gradually made and which I have covered by a considerable number of patents in the United States and foreign countries. They relate to methods and apparata for the production of electrical oscillations and means and methods for transmitting electrical energy without wires. In the last two years my energies have been largely taken up by the construction of commercial plants.

6 Q. Will you refer to some of the lectures or published papers which contain accounts of the experiments referred to in your answer?

A. A collection of my papers up to 1893 has been published in book form by Mr. T. C. Martin, and since that time a number of my articles have appeared in various technical periodicals, as the *Electrical Review* and the *Electrical World.* I may also mention a paper before the Electro-Therapeutic Society, which was published at the eighth annual meeting. More recently I have written an article in the *Century*, giving an account of my work, which partially refers to the subject in question. The date of publication is June 1900. Another article was published in the periodical by Mr. Martin, and appeared in the number for April 1895.

Counsel for Tesla gives notice that at the end of this cause he will produce and refer to the following publications referred to by the witness in his last answer.

The Inventions, Researches and Writings of Nikola Tesla, by Thomas C. Martin, New York, 1894.

The Transactions of the American Electro-Therapeutic Association, Eighth Annual Meeting, published by the association in 1899.

The *Century* magazine for April, 1895, and particularly the article beginning on page 916 entitled "Tesla's Oscillator and Other Inventions" by Thomas C. Martin.

The *Century* magazine for June, 1900, and particularly the article beginning on page 174, by Nikola Tesla.

It is agreed that these publications may be used as a part of the record of this case, although not formally offered in evidence as exhibits.

7 Q. Are the illustrations in the two *Century* articles photographic reproductions of any apparatus which was actually constructed and used by yourself?

A. Yes, sir. They illustrate apparata which I have constructed and frequently used and phenomena produced by them.

8 Q. Please state what is meant by the phrase "tuned circuits," and state somewhat more fully the extent to which you have experimented with such circuits?

A. The word "tuned" is now adopted to designate a circuit having a period of its own which is the same as the period of the oscillation impressed upon it, or which is, as the case may be, a harmonic of the same. In the development of my apparata and methods to which I have made reference, it was impossible to proceed without constantly using such circuits. In fact, the majority of experiments which I performed were only possible by careful adjustments of the periods of vibration. I have frequently had in my laboratory up to a hundred of such circuits, if not more, and have daily used them.

9 Q. Will you please state what led you to a conception of the methods and apparata or combination of apparata which are described in your applications involved in this interference, and referred to in the issues as stated by the Patent Office?

A. In carrying on the work which I have outlined, the tuning of circuits became of necessity a special study. A great many practical problems which I attempted to solve were dependent on the perfection of this art, and I concentrated my efforts to bring it to as high perfection as possible. Exactly when the methods here involved took definite shape in my mind I am unable to state at present on account of absorption in important work in which I am now engaged. But, to my best recollection, it was sometime in 1898 that I realized the invention and began to practice it frequently. This recognition came gradually, and I have not the slightest doubt that long before that if any one would have requested me I might have practiced the invention, for I have an instinctive knowledge of the same. Very often I would connect to an oscillating circuit which emitted a number of vibrations several tuned circuits, each responding to a particular note, and repeatedly I observed that the tuning was extremely sharp. To explain, I will say that a device which I used to indicate the intensity of the oscillations would be practically brought out of action by a very minute change in the constants of the apparata I was using. It did not take me long to find out that this took place when the operation of the device was due to a combined effect of two or more oscillations. I also distinctly recollect that in studying Spencer's works I began to associate tuned circuits with nerve strands, and commenced to perceive vaguely the possibility of a high degree of individualization by a combination of heterogeneous elements, all co-operating in producing result. But, as stated, I can clearly picture in my mind some experiments which I performed in 1898 and which have a direct and intimate bearing on the invention described in my applications or cited in this interference.

10 Q. Did you make an extended trip to Colorado, and if so, when and for what purpose?

A. I left for Colorado, I think, on the 11th of May, 1899, prompted by a number of considerations. One of these was that I had already reached the limits in the power of the oscillatory apparatus in my laboratory, 46 East Houston Street. The largest discharges which I could produce there measured 16 feet from end to end, and that was about as much as was possible to obtain in the room. Another reason was that I wanted to make exact measurements in order to be quite sure that certain rules relative to the propagation of electrical currents through the earth were rigorously correct. Then again, I wanted to be free of the disturbing influences in the city which make it very difficult to tune circuits, of the kind I was chiefly employing, with great precision. I also recognized as a necessity to carry to a still further perfection the invention in interference, and in general to master as completely as possible the art of tuning which I had studied previously in a number of aspects. There were other important reasons which prompted me to go to Colorado in preference to other places which I might have selected, but these bear less on the present subject. On my way there, I delivered a scientific address acquainting some practical business men with a number of scientific ideas on which I was at work and gave an illustration of an art which I had previously originated and which I have designated "Telautomatics." My experiments were made before the Commercial Club and were at the time widely commented upon.

11 Q. What was the date of this lecture before the Commercial Club?

A. The club did me the rare honor of changing the date of its meeting, making it the 13th of May, 1899, on which day I lectured.

12 Q. Prior to your departure for Colorado, on May 11, 1899, had you reduced to practice or practiced the methods and used the apparatus described in your application in interference and referred to in the issues as stated by the Patent Office, and, if so, in what ways and where and to what extent?

A. I practiced the invention almost daily in my laboratory before mentioned prior to my leaving for Colorado, and in the experiments I performed I frequently used apparata almost identical with some of those illustrated in my application. I had developed the invention thoroughly through all its stages, from a simple selective system to a system free from extraneous, willful or accidental disturbance, which was the next step in its perfection. Then to a system which, in combination with these qualities, would also be perfectly private. In fact, I had carried the development to the highest stages, which are the individualization of a message or signal, resulting from a combination of great many characteristics, and finally to what I consider the ultimate stage of perfection, as it appears to me, the individualization in combination with continuous change–according to a predetermined code, key or safety combination–of the signals or messages.

13 Q. Please describe some specific forms of apparatus which you used prior to your trip to Colorado, and state how you used them in carrying out the invention here at issue?

A. One of the most convenient ways was to place circuits in the middle of the room and then excite them by oscillations passed through a large loop or cable laid around the room. As my transmitting apparatus was very powerful I had no difficulty in concentrating a considerable amount of energy — up to half of one-horse power — in one circuit, and this, of course, made it very easy to operate. This method I used very frequently for convenience. Generally, however, I employed circuits arranged as in my system of wireless transmission of energy, which I have described in the patents to which reference was made in my answer to question 5. Each element of the transmitter would be a coil or conductor which had one of its ends connected to the ground, and the other to a metallic plate or conducting body supported in the air. Each element of the receiver or receiving apparatus comprised a similar conductor which had likewise one of its ends connected to the ground and the other, in the same way, to a terminal of capacity. The currents produced in the receiving circuits were sometimes directly utilized to affect the circuit or circuits including the relays or like instruments, or I used on each receiving element a secondary coil and connected the same to the receiving instruments in various ways, so that the operation of the receiver took place only when all, or a certain number, of the receiving circuits which were obviously tuned, responded to the oscillations of as many transmitting circuits.

14 Q. Were the receiving instruments and transmitter always used together in the same room in your laboratory?

A. No, sir; they were not. Sometimes I would take them to the upper floors, or on the roof of the building, or I would carry them to other parts of the city.

15 Q. With reference to the transmitting apparatus which you employed prior to your trip to Colorado, please describe its construction and operation as much in detail as possible, referring if you please to any

patents or publications which will enable the Commissioner of Patents to understand clearly what it was?

A. Referring to the apparatus which I often employed for convenience in practicing the invention, the transmitter comprised a high tension transformer, giving about 50 kilovolts pressure, which charged several sets of condensers. These condensers were automatically discharged so that a number of oscillations, two, three or more, in extremely rapid succession were passed through a cable laid around the room. When I used arrangements of circuits, such as illustrated in my patents on the transmission of energy, I most frequently directed the oscillatory discharges of the condensers through primaries placed in inductive relation to the circuits which were grounded or, generally speaking, connected to the medium which served as a means of transmission, which in some cases was wire.

16 Q. Have you any photographs, or drawings or sketches which will clearly illustrate the construction of such transmitters as you employed prior to going to Colorado, and which were capable of and did actually produce two or more oscillations of different frequencies?

A. Yes, sir; I have drawings, photographs, sketches, models and apparata which I have employed. A very convenient way of producing two oscillations to which I often resorted when I used only two receiving circuits was the following: I employed a large coil in the form of a flat spiral, one end of which was connected to the ground and the other to a conducting body supported in the air. This coil I excited by a primary so proportioned that when the latter was closed by the make-and-break disk which discharged the condensers, the oscillation in the secondary was quickened to much above the rate which the secondary or spiral coil vibrated when the primary was opened. In this way I obtained two oscillations, which I regulated and controlled in a number of ways, and by means of which I excited two tuned circuits, each adjusted to respond to one of the oscillations which were impressed upon the ground in rapid succession by the transmitting coil described.

17 Q. Do I understand that you have this, or similar transmitters, in your possession at the present time?

A. Yes, sir; this very coil in my possession, only since I have left for Long Island the primary, which was wound on a wooden frame and comprised two turns of very heavy cable, was taken down. The coil to which I refer was placed on rollers so that it could be put at any desired distance from the exciting primary, in this way permitting me go graduate the effects and to produce the two vibrations required.

18 Q. What are the dimensions of the apparatus? In other words, give us some definite idea of its size or physical dimensions?

A. The same coil I have used in other experiments, and it was described in one of my patents on the transmission of electrical energy, to which I have repeatedly referred. The patent bears the number 645,576.[1] The question will be found answered in the patent named. I have only to add that the frame upon which the coil was would formed a square of 8 feet.

> Counsel for Tesla gives notice that at the hearing of this cause reference will be made to Nikola Tesla patent No. 645,576, referred to in the last answer of the witness.

19 Q. You have stated that you have some illustrations of the apparatus referred to in the last question and answer. Will you please produce and describe the illustrations, if you have them?

[1] Refer to Appendix B.

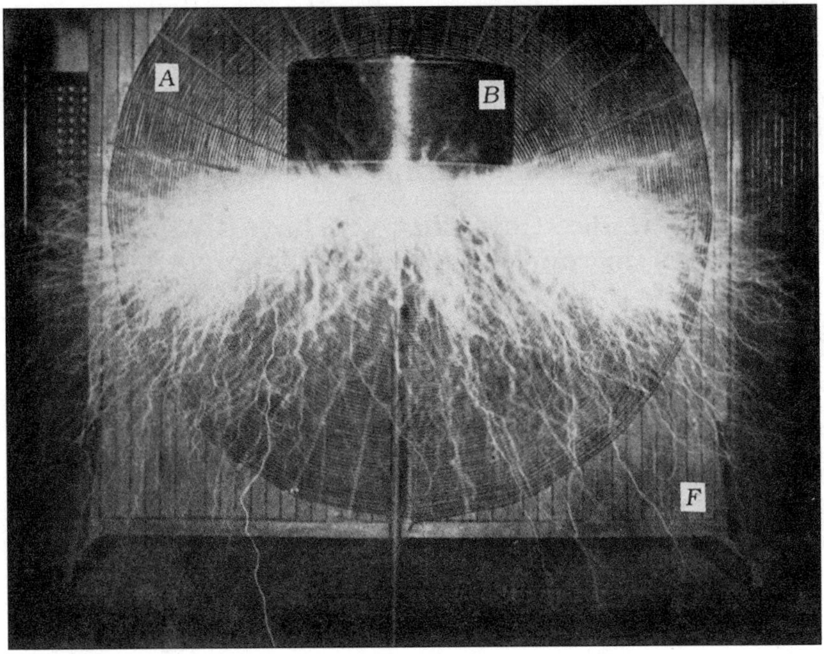

Tesla Illustration, 8-Foot Diameter Transmitter Coil.

A. I have here a cut from the *Electrical Review* of October 26, 1898, in which there is shown in the background a flat spiral coil which I have marked *A*, and which is the coil in question. There is a frame made of wood, which I have marked *F*, and which in reality is 8 by 8 feet, as before stated, and upon which there are wound, upon notched supports, about 50 turns of well insulated cable No. 8. The end of the coil, which is on the outer side of the spiral, was usually connected to the ground, and the end in the center to the insulated conductor of capacity, not shown in the cut. The coil is easily recognizable in the figure, presenting, as it does, the appearance of a spider web. *B* is a coil supported on an insulated stand in front. The adjustment is such, that when the apparatus is excited the maximum potential developed in the circuit falls together with the lower end of coil *B*, from which then powerful streamers issue, which are shown in the cut. I wish to state, however, that the coil marked *B*, was here used to illustrate another experiment and might or might not be there. The primary is not shown here, being behind the frame. To excite the apparatus I used condensers which were charged from the 50 kilovolt transformer and discharged in rapid succession through the primary. The transformer is not visible.

20 Q. Was this transmitter in regular use in your laboratory prior to May, 1899?

A. Yes, sir; it was in regular use for a number of years prior to my going to Colorado, having been constructed, if I remember rightly, sometime in 1895, but it was not used in the manner described to produce two oscillations until a much later date, I think sometime in 1898.

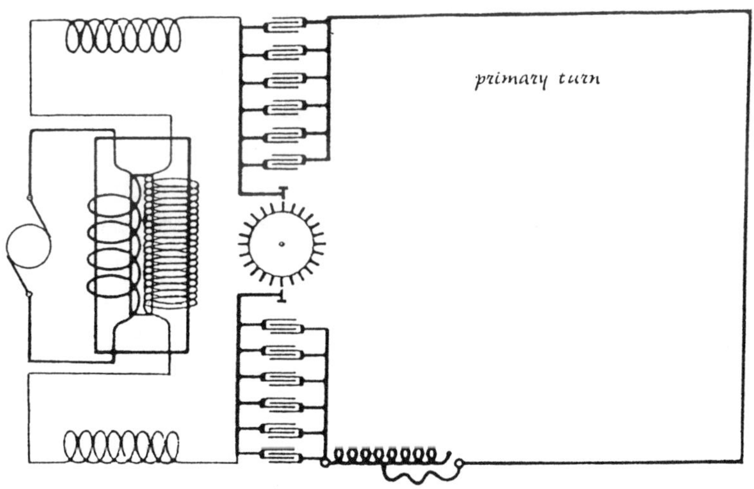

Tesla Exhibit, Transmitter Drawing.

The drawing produced by the witness is offered in evidence and marked "Tesla Exhibit, Transmitter Drawing."

21 Q. Will you please now describe in detail one or more of the forms of receiver which you say you employed with the transmitter referred to in your last answer,

and in which there were two or more elements separately responsive to the oscillations produced by the transmitter?

A. When proceeding in the manner explained in answer to question 16, I would usually employ two coils of a suitable number of turns and would on cylinders of wood of from 1 to 6 inches in diameter and from 2 to 7 feet in length. On one end of these coils I would most frequently wind a smaller coil, either of coarse or of fine wire, according to the character of the circuit which included the receiving devices proper. When I used a sensitive device such as, for example, a resistance of small mass, a few turns were generally sufficient. When, on the other hand, I employed a detector in the nature of a microscopic dielectric film, a great number of turns were provided to obtain a high electromotive force. These small secondary coils, which were wound, as a rule, on those ends of the receiving coils which were connected to the ground, and through which the strongest currents passed, were then associated or connected in the circuit in many ways, so that the receiver would be operated only when both of them were energized and would fail to operate when only one of them was energized by the oscillations of the transmitter. It will be understood, of course, that the oscillations of the transmitter were passed into the ground, through which they spread and, producing variations of potential at the place or places at which the receiving coils were grounded, excited the latter. The currents thus circulated in the receiving coils energized the secondaries placed in inductive relation, and the secondary currents in turn affected the sensitive devices or detectors which then closed a local battery circuit, thus bringing into action a relay or other device. Sometimes I simply combined the two secondaries through a common path and operated by the current passed through them a device such as a small mass resistance, the variations of which I then again utilized to cause the actuation of some instrument in various ways.

22 Q. What were some of the instruments that you operated by the combined action or effects of two or more oscillations, such as are now under consideration?

A. Most generally they were telegraphic relays or telephonic receivers, or clockworks utilized by me for specific purposes. I also operated what I have termed a "Telautomaton," which was in the shape of a boat provided with motive machinery, steering mechanism and various appliances for performing a great variety of operations. This boat, or at least one of the boats I constructed, I described in my article in the *Century*, of June, 1900, which contains a photograph of the machine. Originally the boat was actuated with one single circuit, but very often I operated it by two tuned circuits and two vibrations. This boat will also be found described in a patent granted to me November 8, 1898, which bears the number 613,809.[2] The drawings in this patent were made from the actual machine.

> Counsel for Tesla gives notice that at the hearing of this cause he will refer to the above patent of Nikola Tesla, No. 613,809.

23 Q. Can you describe any other forms of receiver which you used in practicing this invention prior to your Colorado trip?

A. Yes, sir; I can describe a number of such instruments I used. One in particular, of which I have already spoken, I believe. This was a coil wound around a cylinder of large diameter covered on top with a board so that it resembled a table on which I often placed the apparatus for receiving. Within this coil was another

[2] Refer to Appendix A.

NOTE: The illustrations on this page and accompanying commentary are not a part of the formal transcript of interference depositions.

Tesla's response to question 23 refers to Figures 9 and 10 of the article, "Tesla's Oscillator and Other Inventions," by T. C. Martin, in the April 1895 issue of the *Century* at pages 926 and 927, respectively. These figures are also referred to later in the transcript, and are reproduced here together with the original accompanying captions for purpose of reference.

Fig. 9. Experiment showing play of electric sparks between condenser plates, produced by electric charge. The coil, standing in the center of a large room, is unconnected with the energizing circuit. (From flash-light photograph).

Fig. 10. Experiment showing the lighting up of an ordinary incandescent lamp, at a distance, through the influence of electrified ether-waves. (From flash-light photograph).

coil, and the terminals of both were led to the board so as to be readily accessible. I very frequently used this apparatus in the following way: I would pass through the cable laid around the room two electrical oscillations of different periods and would tune one of the coils mentioned to one of the oscillations, and the other to the other oscillation. I used frequently condensers and inductances for this purpose. As stated before, this apparatus was very convenient to use because I was familiar with the vibrations, the constants of the circuits, and all other particulars, and could complete the test in a very short time. To give an idea I may mention the article in the *Century* of April, 1895, by T. C. Martin, in which such an instrument is shown in Figures 9 and 10. This apparatus illustrated in the *Century* article was, however, destroyed by fire, but in my laboratory in Houston Street I constructed a similar one, which I have frequently used as stated. Besides this, I had a great many other apparata of this kind which I frequently employed in my demonstrations or lectures with tuned circuits. I may mention, for instance, a coil which was provided with a handle to which one of its ends was attached, the other end being connected to a loop of rather thick aluminum wire, which served as capacity terminal. On the lower end of the coil, near the handle, was wound another coil of few turns of wire, to which I often connected a small condenser, adjusting its capacity so that the coil was powerfully excited when a certain vibration was passed through the cable around the room, the other coil responding to another vibration, properly chosen, so that both vibrations would not greatly interfere with each other. It was a common experiment with me to take a hoop of wire closed through a minute lamp and light the latter by the conjoint action of the two coils. In fact, I was compelled to adopt this expedient because frequently in my demonstrations, when I showed my experiments to an audience, it would happen that when

I wanted to operate single circuits with some devices more than one circuit would respond, and people would call my attention to this, and I would then have to say that the circuits were not carefully tuned. But, by adopting two vibrations through the cable around the room I could always manage to operate one single device out of a great many. Besides this coil, which I frequently used, I had probably a dozen of others, similarly constructed and adjusted, which I showed on many occasions.

24 Q. Please describe the details of the construction of this hoop and lamp, and the manner in which the lamp was operated by the conjoint effect of the two oscillations.

A. I will give a description of the apparatus as closely as I am able to do now from memory. I think the coil, which was wound on the outer side of the frame of cardboard, about 30 inches in diameter and 5 inches wide, had something like 300 turns of wire — I should say No. 20. One end of this wire was connected, as I stated before, to a metallic handle, and the other to an aluminum wire, fastened on the end of the frame opposite to the handle, so as to form a circle, which, however, was not closed. This aluminum wire, constituting what I have called the capacity terminal, was bare, because when the action was powerful I would get strong streamers from it. On the inner side of the cardboard frame were wound several turns of thicker wire, and these were connected, in the experiment, to a small condenser which was so proportioned that the period of oscillation of the circuit connected to it was suitable for the test, and different from the period of the other circuit which I have first mentioned. The transmitting apparatus most frequently used was already described in answer to question 15. It comprised two or more sets of condensers, which were charged from the transformer giving a pressure of 50,000 volts and discharged, in most cases alternately and in very rapid succession, through the cable around the room. I would then take a loop of wire forming a circle, to which a device such as a minute lamp requiring an extremely small amount of energy was connected, and would approach the loop to the active system of the two circuits, energized in extremely rapid succession from the cable around the room, and the lamp would be lighted; or, another device which I used, operated by the combined action of the two circuits, while other tuned circuits which were in the room would not be perceptibly influenced. I refer to this experiment simply to illustrate one of the many ways in which I illustrated the invention in my laboratory from 1898 until I moved to Wardenclyffe, recently, in the spring.

25 Q. Please explain with reference to the last experiment to what cause the current in the loop, including the lamp, was due, and in what way the combined action of the two coils was necessary to the operation of the lamp?

A. The loop containing the lamp was, of course, excited by both circuits in rapid succession, but the minute filament had no time to cool down, so that the action of the two circuits on the loop, and consequently on the lamp, or other device connected to it, was practically simultaneous and cumulative. In other words, the lamp was lighted by the conjoint action of both the circuits, whereas it would not have been lighted by a single one, at least not the way I adjusted the conditions.

26 Q. Were any of the operations which you have described in your preceding answers as having been performed by you in and about your laboratory prior to your departure for Colorado witnessed by others than yourself?

A. Yes, sir; they were witnessed by a great number of people.

27 Q. How many of these operations which you have described were seen by others at that time?

A. Most all of them. As a matter of fact, I would make it a rule, in order to entertain my visitors and to illustrate the progress I was making, to vary the experiments as much as practicable. The necessity of operating as many circuits or devices as possible without interfering with the other ones compelled me to adopt the methods now under consideration, and by their use I was enabled to make my demonstrations more successful than I could have made them in using simple tuning or single circuits.

28 Q. What assistance had you at that time in your laboratory?

A. Usually, when I was making such demonstrations, my secretary, Mr. George Scherff, was present and aided me; and, on various occasions, my draughtsman, Mr. Diaz Buitrago. I also had one or two boys to assist me in preparing the experiments and in showing them.

29 Q. Prior to your Colorado trip, in May 1899, had you described the inventions or improvements of your application now in interference and referred to in the issues to others, and, if so, to whom?

A. Yes sir; I explained them to Mr. Scherff, and I began, before I went to Colorado, to prepare an application, but had to abandon this work for want of time. Mr. Scherff himself urged me very strongly to place in the Patent Office some kind of document protecting me. I remember this distinctly. I was unable to carry out this purpose until I returned, sometime in January of 1900, if I rightly recollect, and owing to pressing duties, the filing of the document was postponed very long, I believe until July of the same year.

30 Q. State what you disclosed to Mr. Scherff that has a direct bearing on the specific purposes of the invention of your application now in interference. I refer now only to disclosures made prior to your Colorado trip?

A. I explained to him distinctly that by combining two or more differently tuned, or, speaking broadly, electrically different circuits, each affected by a vibration of the transmitting apparatus, I could secure a privacy of messages and could send through the earth, or, as the case may be, through a wire, a great many messages simultaneously and also establish a safe and sure communication between points in spite of all accidental or willful disturbing influences. I think he understood perfectly well that a receiving apparatus operated by the conjoint action or effect of a number of tuned or electrically different circuits was incomparatively safer against disturbance than a circuit or receiving apparatus dependent for operation on a single vibration. He understood why, by the use of this principle, a much greater number of messages could be sent simultaneously through a common channel, and he appreciated fully the commercial and practical value of the method. I think he also comprehended the higher refinements to which I have already referred, and which aim at individualization of a message comparable in degrees to that of a safety or combination lock, and which is attained by making a great number of distinctive elements cooperate in causing the action of the receiver.

31 Q. Had Mr. Scherff seen you performing these various operations before you disclosed to him the specific purpose of your applications in interference?

A. Yes, sir; he has seen me perform these experiments for a long time, possible for a year or so, before I thought of communicating this present invention to him. He was employed by me, originally as stenographer and typewriter, and one of his duties was to follow the publications and call my attention to matters of importance. He surprised me by his quick comprehension of the principles, and when I discovered this I began to explain to him more in detail what I was doing. The exact date when this invention was communicated to him I cannot remember. I know only for sure that it was sometime before I went to Colorado to carry my work in these directions.

32 Q. Did Mr. Scherff go to Colorado with you?

A. No. sir; he remained at my laboratory, 46 East Houston Street, New York City.

33 Q. And has he been since that time in your employ?

A. Yes, sir; always.

34 Q. Please state now, in a general way, the nature of your work in Colorado; stating the character of apparatus which you constructed, and the general nature of the work which you carried on there?

A. I went to Colorado Springs in preference to some other place in Colorado for various reasons which are here immaterial, and I will mention only one which I think necessary. I had discovered some time before the surprising fact that air, when subjected to tensions of several million volts producible with oscillatory apparatus, became conductive so that I could pass currents of several hundreds of amperes through it. Experiments and measurements convinced me that it was a mere question of mastering the difficulties in producing currents of sufficiently high tension and maintaining the terminals at a certain elevation in the atmosphere, and energy in large amounts could be transmitted without wire to great distances. I observed also certain influences of elevation on the capacity of the terminals I used in my circuits. Colorado offered me the best opportunities for continuing my observations and bringing them eventually to a practical fruition. The predominating idea, however, was to produce an apparatus which would be capable of disturbing the electrical condition, if not of the entire globe, or at least of a very extended portion of the same, thus enabling me to transmit intelligence to great distances without wire. Again, in regard to the tuning of circuits, I required a place remote from disturbing influences and preferably elevated to study more precisely the forces active in disturbing the potential at any point of the earth's surface. Now, for all this I needed a powerful plant and, as my system of transmission by alternating currents was practiced on an extended scale in Colorado, I felt sure that I would get all the help I could reasonably expect. I was received as I anticipated, and the El Paso Electric Company, through the intervention of my friend, Leonard E. Curtis, offered me power and every other assistance most liberally. As soon as the laboratory was erected the wires were run from the station a few miles away, and I was able to draw, according to necessity, from their circuit 100 horse power and more. I immediately began to carry out my plan which I had, in a large measure, already decided upon in New York in regard to the construction of a powerful oscillator on the principle originated by me, which in now well known. I encountered considerable difficulties which were fortunately slowly overcome, and sometime in August I was already able to produce electrical discharges of about 18 feet length. Having had in mind to perfect still farther the methods and apparata involved in this interference, I designed my oscillator so that it could emit normally two and, if necessary, more oscillations. I may describe a little more in detail the arrangement first adopted. I employed a Westinghouse transformer which was rated at 50 kilowatts, but which, I found, could stand easily three times that output. It gave 60,000 volts when normally connected, but as this tension was found unsuitable I made a change and operated with 20,000 or 40,000 volts. From this transformer I charged two sets of condensers. These were of a kind patented by me some time ago in the United States. The dielectric was glass and the conducting coatings were formed by an electrolyte which, in long experience, I have ascertained to be most efficient. These two sets of condensers were connected, either each to one separate primary or they would both discharge, alternately or simultaneously, through one common primary. In the early stages of experiment I used to tune the two circuits so that they gave the same vibration, but when necessary I altered the vibration of either or both of them so as to get two different vibrations. In the first oscillator the two primaries, or as the case might be the one common primary, were placed on wooden supports around the laboratory in a circle of about 15 meters in diameter. One of the first transmitting coils was wound on a wooden frame forming a similar circle and had, I think, something like 30 turns. To the top end of this transmitting coil I would connect the capacity terminal which was supported above the laboratory, and the other end near the primary I would connect to earth either directly or through a condenser properly adjusted. The general rules which I have given in my patents, first referred to in answer to question 5, were carefully observed, and when the apparatus was in action the maximum potential developed in the transmitting circuit was made to fall together with the elevated terminal, and thus a current of many hundreds of amperes was made to flow in this circuit when the primaries were energized, passing away into the earth and creating a disturbance. I would measure the effects produced and calculate the distance at which a certain amount of energy, necessary for the operation of a receiving device, would be transmitted. I would remedy the defects and improve, and so until I finally succeeded in producing an apparatus capable of doing what I expected it to do.

35 Q. Your work in Colorado was in the nature of investigation rather than an effort to establish a practical or commercial plant, was it not?

A. No, sir; I cannot say that. I carried on this work with practical ends in view, and whatever scientific investigation I made it was simply a means to this end, although I sought certain lines of experiment, when the time permitted, which were of an interest, purely scientific, at that time.

36 Q. You probably misunderstood my question. Was it your object to establish in Colorado a commercial plant for the transmission of messages?

A. Not exactly that. I desired to ascertain the best conditions under which a power plant could be utilized for the transmission of messages across the Atlantic and Pacific Oceans, of the practicability of which I had already gained the conviction before I went away. I also intended to perfect all the details of my system before embarking into a commercial enterprise. In other words, although as far as I myself was concerned I had made perfectly sure, by experiments and measurements in New York, that I could establish such a communication, I wanted to be certain that not only could I transmit the messages but that my undertaking would be a commercial success.

37 Q. Do you recall when you returned to New York from Colorado?

A. Yes, sir; I think it was in the middle of January, 1900.

38 Q. While you were in Colorado did you devote any time and attention to the particular matter of this interference?

A. Yes sir; I believe I stated that. My apparatus was in fact originally designed for practicing the methods involved in this interference.

39 Q. Did you continue the experiments in Colorado which involved this invention; and, if so, in what way?

A. Yes; I did. I never lost sight for a moment that, next to perfecting apparatus capable of producing oscillations of that immense power which I was attempting to attain, it was most important to perfect methods and apparata for utilizing the effects for the purposes of telegraphy, or generally speaking, the transmission of intelligence. I may describe one or two ways in which I used the apparatus in perfecting the invention. Sometimes I would utilize the two condenser sets to produce, by means of the break disk, alternating discharges in two independent circuits, with which I would energize either two circuits connected to the ground, or, as the case may be, one grounded circuit. I would thus obtain two different oscillations, to which I would tune two receiving circuits, combining them in the manner described in my applications here involved, or in other ways. In other instances I would produce two different oscillations for similar purposes by means of a single transmitting circuit in the manner which I have already described, in answer to question 16, using but one primary. At other times I would employ the transmitting circuit which was wound on the frame running around the laboratory as one element of the transmitter, and would cause this circuit to discharge into another circuit of a different period, so that two vibrations were obtained in the ground, and two distinct tuned circuits, grounded at another point, were operated and combined in ways conforming with the spirit of the invention.

Again, I can recall that sometimes I would produce more than two oscillations by methods which, though broadly within the scope of the present invention, were different in some particulars.

40 Q. Did you employ any receiving instruments, the operation of which depended upon the conjoint action of these different oscillations; I refer to the time while you were in Colorado?

A. Yes, sir; I did frequently.

41 Q. How were they constructed and in what manner were they used?

A. They were either specially constructed relays, or commercial instruments which I had bought and merely adapted for the purpose. In the latter part of my work in Colorado I used some which were novel and more sensitive and precise in their action. I would connect the instruments usually in an operating room distant from the ground connection of the transmitter, complete the tuning and adjustment, observe the effects and magnify them, if necessary, by certain methods so as to enable me to operate with greater precision, then begin reconstruction and so over and over again.

42 Q. What assistants did you have in your work in Colorado?

A. I had first of all a German engineer, Mr. Fritz Lowenstein, who came to me sometime in April, 1899, just as I was making preparations to go to Colorado.

43 Q. How long did he remain with you in Colorado?

A. To my great regret, he left my employ, I think, towards the end of September, 1899, just about when

I was beginning to really want him. Family affairs compelled him to return to his home in Germany.

44 Q. He came back to your employ subsequently, did he not?

A. Yes, sir. He was greatly impressed with my work, particularly with what he saw in Colorado, and just as soon as his family affairs were arranged he thought of coming to America again. If I rightly recollect, he presented himself at my office in the beginning of February of this year,[3] and I received him gladly. I have since employed him on the details of my plant at Wardenclyffe, Long Island.

45 Q. Have you any memoranda, in the shape of sketches, or descriptions of the invention of your application here in interference, which were made prior to your return from Colorado?

A. Yes, sir, I have. I have with me here my notes for the month of June, 1899, which I shall cheerfully submit for inspection. There are a few sheets bearing an intimate relation to the invention under discussion, and I will particularly call attention to the sheet which I wrote in Colorado Springs, June 27, 1899.

> Counsel for Tesla offers in evidence the sheet referred to by the witness, and the same is marked "Tesla Exhibit, Colorado Sketch." The remaining sheets produced by the witness are offered for the inspection of opposing counsel.

It is stipulated and agreed that photolithographic copies of the sheet offered in evidence may be used in lieu of the original, it being understood that the original will be produced at any time when it may be called for by counsel for Fessenden.

Adjourned to Wednesday, August 6, 1902 at 10 A.M.

---

[3] 1902.

New York, August 6, 1902.

Met pursuant to adjournment.

Present — Counsel as before.

46 Q. Please read to the Examiner the contents of the "Tesla Exhibit Colorado Sketch"?

A. "Colorado Springs, June 27, 1899.

"Arrangements of apparatus in telegraphy through the natural media aiming at exclusion of messages in accordance with matters experimented with in New York. This is not quite so good as method with condenser of commutating individual impulses, but a great safety can be secured nevertheless. The idea was to provide more than one synchronized circuit and to make the receiver in its operation dependent on more than one such circuit. Experiments have shown that a great degree of safety is reached with two circuits. I think with three it is almost impossible to disturb the receiver when the vibrations have no common harmonics very near to the fundamental tones. Several arrangements experimented with are illustrated below. These are to be followed up. Figures 1, 2 and 3 illustrate some arrange-

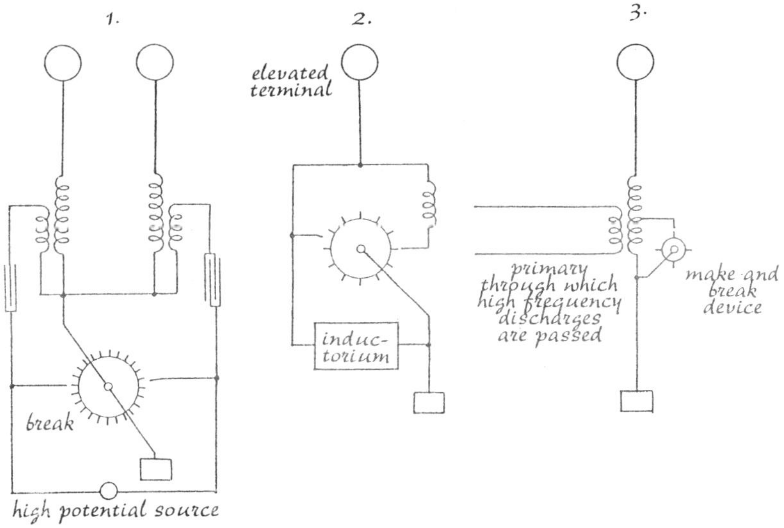

[For explanation of "inductorium," refer to Appendix F.]

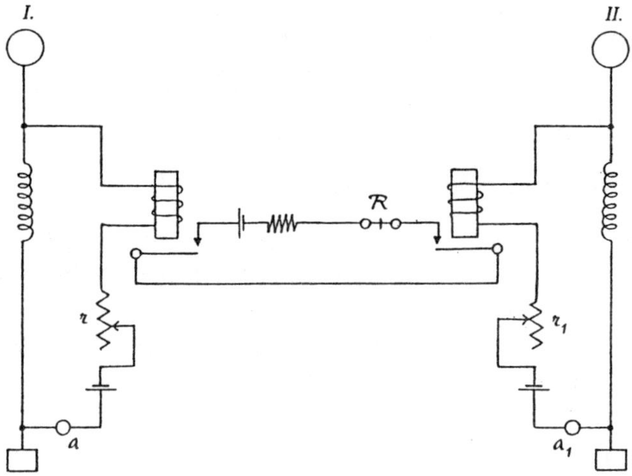

ment of apparatus on the sending station, by means of which two vibrations of different pitch are attained. A great number is omitted to be shown for the sake of simplicity. In case one two sending circuits which should be some distance apart are provided and energized alternately by discharging condensers of suitable capacity through the corresponding primaries. In the diagram 2 and 3, one sending circuit is arranged so that its period is altered by inserting some inductance, as in 2, or by short-circuiting a part of circuit periodically by means of an automatic device. Such a device is not necessary to use, however; arrangements of this kind will be later illustrated. On the receiving station two synchronized circuits, responding to the vibrations, each to one of the sender. The receiver *R* responds only when both circuits I and II affect sensitive devices a a_1. Diagrams are self explanatory."[4]

[4] These diagrams correspond to those presented in *Nikola Tesla: Colorado Springs Notes, 1899-1900* (Belgrade: Nolit, 1978) p. 48 (June 27, 1899 entry). Some inconsequential differences can be noted in the text for the entry as presented in that published work and as reported in this deposition from Tesla's oral reading of them.

47 Q. What was your purpose in writing the above description?

A. As nearly as I recollect now, I was preparing to systematize my work in these directions, and this was the first sheet, if I remember rightly, which I penned and proposed to use later on in working out all the details systematically and covering the whole ground. I do this as a rule.

48 Q. What are the other sheets which you have produced, together with the above exhibit?

A. They are similar notes which I put down for the same purpose as the work presented itself. Many of them are, I think, simply calculations or arrangements of apparatus or data of some kind.

49 Q. When did you make the "Tesla Exhibit, Colorado Sketch"?

A. On the 27th of June, 1899, as marked on the top of the page.

50 Q. When did you write that data there?

A. When I wrote this sheet.

51 Q. When was this sketch first shown and described to others?

A. In making notes I would sometimes call to my desk my assistant, Mr. Lowenstein, and it is just possible that as I was making these sketches he was present and may have seen them. But of this I have no knowledge. I recollect distinctly, however, that on coming back from Colorado Springs in January, 1900, I took all my notes, sat down on the desk, and turned leaf after leaf before my secretary, Mr. Scherff, telling him of the various things I had done and that he would, in a general way, recollect my showing it to him.

52 Q. How soon after your return from Colorado was this done?

A. This was done, I believe, on the first day I came. I have recorded in these notes two or three wonderful discoveries which I made, and I considered them so valuable that I would not entrust them to the express but carried them in a hand bag which I brought to the laboratory in New York.

53 Q. Have you any other written description of the invention of your application in interference, and if so will you please produce it, stating when and under what circumstances it was prepared?

A. Among my papers I also brought the text of the larger part of the application, which I have since filed, and which is involved in this interference. When I began to write the text, I cannot tell exactly, but I think it must have been some time in October or November of 1899. I know that I wrote it with great care, and as far as the larger part of the application up to the detailed description of the apparatus is concerned, it was scarcely changed at all. Only a few days after my coming to New York in January, 1900, I took out the text which I had written, called my secretary and began, as I usually do, to prepare the documents which I intended to file. I dictated to him the latter part of the application and he typed it. I thereupon corrected the copy and he typed it over again, and so probably a dozen times, as I invariably do, sometimes changing merely a word. The *Century* began then to press me very hard for completing the article which I had promised to them, and the text of this article required all my energies. I knew that the article would pass into history as I brought, for the first time, results before the world which were far beyond anything that was attempted before, either by myself or others. The work on this article almost unnerved me, and it took some time before I could concentrate my thoughts on the application which was almost completed at that time. I had, however, a still more pressing discovery to record in the Patent Office, a description of which I had also brought from Colorado, and I thought it best to first file the other application, which was done on May 16, 1900. Soon after I finished the text and brought it to my attorneys who made insignificant alterations and drew the claims.

I herewith produce a type copy of the application made by my secretary at about that time, who will, I think, have no difficulty in recognizing in the first part of this document my original text. When this copy was made I cannot tell. It is very likely that it was at about the time when the application went to file, or a little earlier.

> The document referred to by witness in his last answer is offered in evidence and marked "Tesla Exhibit Draft of Specification."[5]

54 Q. Please state what, if anything, you did after returning from Colorado in the further development of the invention of your application in interference?

A. On my return I continued to work out all the details of commercial plants in which I was to apply this invention in a still higher form of perfection. These plans, designs, and calculations took four or five months, and when they were completed and most of the important particulars settled upon, I sought financial support for carrying out my undertaking of establishing communication without wires across the seas. The support was most generously given, and since that time I have done all I could to complete the transmitting and receiving

[5] Refer to Appendix E for the text of the resulting Tesla U.S. Patent No. 787,412 granted April 18, 1905.

plants and apparata as soon as possible. My first plans contemplated communication across the Atlantic Ocean, but continuous improvements which I made enabled me to design apparatus competent for the establishment of telegraphic communication to any point of the globe, about a year ago, a slight modification in my plans was made which were, since then, faithfully followed with the exception of a few insignificant details. This work is now far progressed, and I am hopeful to inaugurate a practical and commercial system of intelligence transmission without wires to any desired distance very soon.

55 Q. Do I understand that the plant referred to is in process of construction?

A. Yes, sir, it is still in process of construction, although the power plant and a large part of the appliances which I intend to use are ready.

56 Q. Will you state in a general way what the capacity or size of the plant is?

A. I have an engine of the Westinghouse type rated at 300 horsepower and a directly connected dynamo of 200 kilowatt capacity. There is an auxiliary small engine of about 40 horsepower and a dynamo connected with the same. Also, an air compressing plant and other accessories. Underground conduits lead from the main building to a tower about 250 feet away and of a height of 154 feet from the ground, on which a terminal will be mounted which is to be energized to a high potential in the manner described in my patents on the transmission of energy without wires repeatedly mentioned. The power plant will be operated by two Babcock and Wilcox boilers of 200 horsepower each. The plant is designed to enable me to operate continuously, using the boilers alternately. Normally, sending messages at the rate of 30 words per minute, I shall require 100 horsepower to reach with my apparatus to those points of the globe which require the greatest effort of the transmitting plant. The power of the generator will be transformed in high tension transformers peculiarly arranged and used to store energy in condensers of special construction, and the vibratory energy of the condensers will be employed to excite the transmitting circuit or circuits. By certain novel processes, the energy of 100 horsepower will be used so as to deliver, into the earth, power at a rate of from 5 to 10 millions of horsepower.

57 Q. This is the plant which you are now erecting at Wardenclyffe, Long Island?

The question is objected to by the attorney for Fessenden as irrelevant and immaterial to the question of priority. The objection will be continued without further notice to further questions and answers in the same line.

A. Yes.

58 Q. Was and is this plant designed for operation in accordance with the invention of your application here in interference?

A. Yes, sir; only the devices which will be used are carried to a higher degree of perfection and may, in some details, differ from those illustrated in my application.

59 Q. Can you state in round numbers the amount which you actually expended up to the time of leaving Colorado in the development of the system of which the invention of your present application forms a part?

A. I could get a fairly close estimate by consulting the books. Speaking from memory I should say that about $100,000 dollars were expended up to my return from Colorado, and since that time approximately $150,000.

60 Q. What proportion of this amount do you think was

expended in the development and construction of transmitters for producing two or more oscillations and in the construction and use of receiving instruments, the operation of which depended upon the conjoint action of two or more such oscillations?

A. It would by very difficult for me to tell approximately how much was expended for the transmitters themselves, but I can estimate in a general way the amounts expended. By far the largest expense was incurred in producing the transmitters, and in particular in constructing transmitters of great power. As I look over the work now, I can unhesitatingly state that the expenditures involved in the construction and perfection of the receiving apparata could not have been one-fiftieth of those met with in the construction and development of powerful transmitters such as I have used in Colorado and am to use in my present undertaking. There are innumerable devices, some of them ready in the market, which can be adopted in carrying this invention out, to serve the purposes of the receiving station. Furthermore, it is not necessary to erect large and expensive structures and use costly machinery such as is required at the transmitting plant.

61 Q. Have you ever concerned yourself with the transmission of messages either by the system or your application in interference, or by any other, on a small and insignificant scale, by which I mean, the transmission of messages over distances of a few miles?

A. In the early period of my work, some time in 1895, and possibly in 1896, I did, and in fact I contemplated for some time to introduce a system on a limited scale. But about that time I made some discoveries which enable me to determine and calculate with great precision the effects produced at a distance and which showed that, if I could only evolve a transmitter of adequate power, I could, with my system, bridge the greatest terrestrial distances. From this moment on, the dominating idea became to perfect such a transmitter, and as I made progress I always calculated and kept before me the performance of the apparatus which I was improving until I reached, as I stated before, the limit of power obtainable in my laboratory at Houston Street, which was a rate, I think, of something like 5,000 horsepower. In Colorado, I pushed the rate up to over 100,000 horsepower, and my present apparatus is designed to give a rate of over 5-million horsepower. I may say that, with the facts with which I am familiar, and which I have thoroughly ascertained many times, I am able to calculate exactly the effect which I will receive in any circuit placed anywhere on the globe. These facts I intend to make public shortly.

62 Q. Will you produce photographs of some of the apparatus which you have constructed and actually used in practicing the invention of your application in this interference, and describe them as you offer them as exhibits?

A. I have some photographs which have been forwarded to me from Wardenclyffe yesterday at the request of Mr. Page. They refer partially or entirely to the subject, and I will describe them briefly.

To begin with, this photograph, which is marked "Tesla Exhibit, Photograph No. 1," is the high tension

NOTE: The photograph of this high voltage power distribution transformer was not included with the deposition transcript. However, this transformer is also shown in "Tesla Exhibit, Photograph No. 2," and referenced in the description on page 48 accompanying "Tesla Exhibit, Photograph No. 10."

Tesla Exhibit, Photograph No. 1.

transformer, giving 50 kilovolts pressure, to which I have referred in my answer to question 15, and which was used to charge several sets of condensers. This transformer was taken to Colorado and in some preliminary experiments I used it there.

The photograph produced by the witness is offered in evidence and marked "Tesla Exhibit Photograph No. 1".

Tesla Exhibit, Photograph No. 2.

This photograph, "Tesla Exhibit, Photograph No. 2," was taken in Colorado, and shows two sets of condensers, each enclosed in a box of mahogany and adjustable by plugs from above. These were condensers which were permanently set up in my laboratory, which I frequently used in these experiments and which I also took with me to Colorado. They were employed in some of the preliminary experiments without any change, but subsequently, as I increased the pressure, I made some alterations in them. Just between these two condensers was a box containing the make-and-break disk, which

I have mentioned in my answer to question 16, and which was driven, first by a direct-current motor when I came to Colorado, and subsequently by an alternating-current motor. The disk had, I think, 25 teeth or projections, and on each side of the box containing it there was an insulated terminal, which I used to connect to one of the two circuits when I performed the experiments described.

The photograph produced by the witness is offered in evidence and marked "Tesla Exhibit Photograph No. 2."

Tesla Exhibit, Photograph No. 3.

This photograph, which was marked "Tesla Exhibit Photograph No. 3" is one of the original photographs of one of the automata which I have constructed and operated by the use of two tuned circuits as described in my application. I did this in two ways: either by placing, in the inside of the boat, two circuits which I tuned and combined so as to cause the operation of the controlling mechanism when both of these circuits were energized by their respective vibrations which were passed through the cables around the room, to which I have made reference a number of times; or, in another way, by using two coils and connecting one end of each to the ground and the other to a metal plate or a bundle of wires, and then exciting them, either by means of the large coil in the form of a spiral, which I have described, or by two circuits which were improvised or adopted for the experiment.

> The photograph produced by the witness is offered in evidence and is marked "Tesla Exhibit, Photograph No. 3."

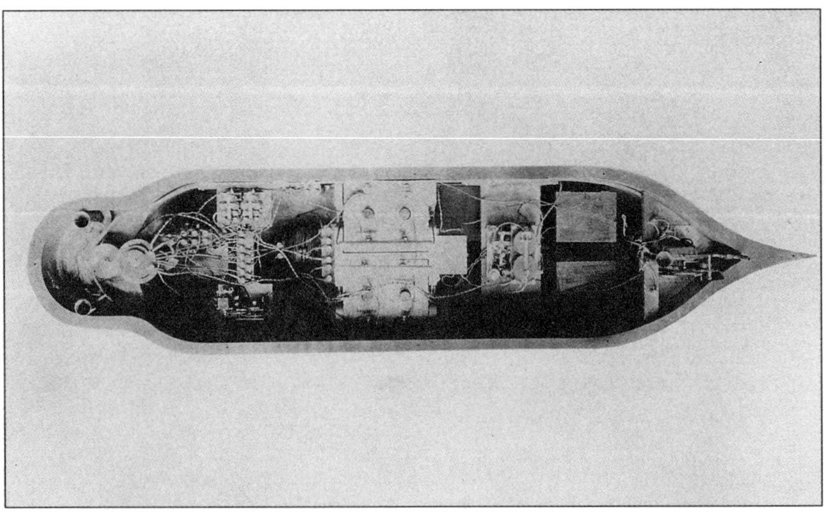

Tesla Exhibit, Photograph No. 4.

Here is a photograph showing another automaton with the cover removed, and the manner in which the receiving circuits were placed when I operated it by the oscillations of two frequencies passed through the cable around the room. I cannot recollect whether the photograph was taken with one or two such circuits, as it is not distinct enough. It is merely offered to show the arrangement of the receiving circuits, and I lay no stress on whether there are one or two such circuits illustrated.

> The photograph produced by the witness is offered in evidence and marked "Tesla Exhibit, Photograph No. 4."

This picture shows, or is a general view of, a part of my laboratory at 46 East Houston Street, New York City, and shows several apparata which I have used in practicing this invention, and to which I have referred. The spiral coil before described would on a square 8 by 8 feet, made of wood, is shown in the background. Looking at it, to the left will be seen a square frame of wood in which there were contained two turns of heavy cable, which I could connect either in series or multiple arc. Those primaries, or respectively primary, were used when I desired to obtain two oscillations in the manner explained in my answer to . question 16. The spiral coil could be approached more or less to the frame, being placed on rollers for convenience, and thus I would graduate the mutual induction so that two oscillations resulted; one, considerably quicker, during the time interval when the primary or exciting circuit was closed, and another, slower one, during the interval when the primary was opened by the make-and-break disk, but both following each other in rapid succession so as to be practically simultaneous as far as the operation of the receiver is concerned. Farther to the left may be observed some condensers which I used in addition to those already described. In front of the picture will be seen the receiving apparatus in the form of a table which had two circuits, the terminals of which were connected to binding posts or studs. This is the receiving apparatus which I have described in my answer to question 23.

> The photograph produced by the witness is offered in evidence and marked “Tesla Exhibit, Photograph No. 5.

This photograph was taken at the occasion of Lord Kelvin’s visit to my laboratory, about two-and-a-half years ago, at the time of the meeting of the British

Tesla Exhibit, Photograph No. 6.

Association at Toronto. To entertain him I used this receiving instrument in a somewhat different way, and the photograph is introduced merely because it shows very clearly every part of the same. It will be noticed that there is a spark passing between two terminals. These are the terminals of one of the circuits to which I have referred several times. A little behind will be seen an incandescent lamp connected to two binding posts. The latter are the terminals of the other circuit.

The photograph produced by the witness is offered in evidence and marked "Tesla Exhibit, Photograph No. 6."

Here are two photographs of two forms of oscillatory apparatus which I often employed to produce two oscillations when sending them over a wire to excite two circuits of different periods and to operate by them a receiver in the same or similar way as illustrated in my application. I wish to state distinctly that these apparata were not originally designed for this purpose, but I merely used them at a certain time in my earlier experiments.

Tesla Exhibit, Photograph No. 7.

Tesla Exhibit, Photograph No. 8.

The photographs produced by the witness are offered in evidence and marked "Tesla Exhibits, Photographs Number 7 and 8," respectively.

In addition to the photographs of these apparata, I may produce here a working drawing of one showing some details, more particularly the two condensers which were contained in a wooden box supporting a motor which drove a make-and-break device. I would usually employ two brushes, one for each circuit, and a common return from the base. The drawing is offered merely to bring out clearly some details. I recollect that in using such apparata as I had on hand I experienced the necessity of providing some means which would enable me to produce a greater number of vibrations and to investigate more systematically the best relations to be observed in transmitting such vibrations over wires. I had an apparatus at about that time and adapted it for this purpose, and I offer herewith some of the first drawings of it which were made, I think at a much

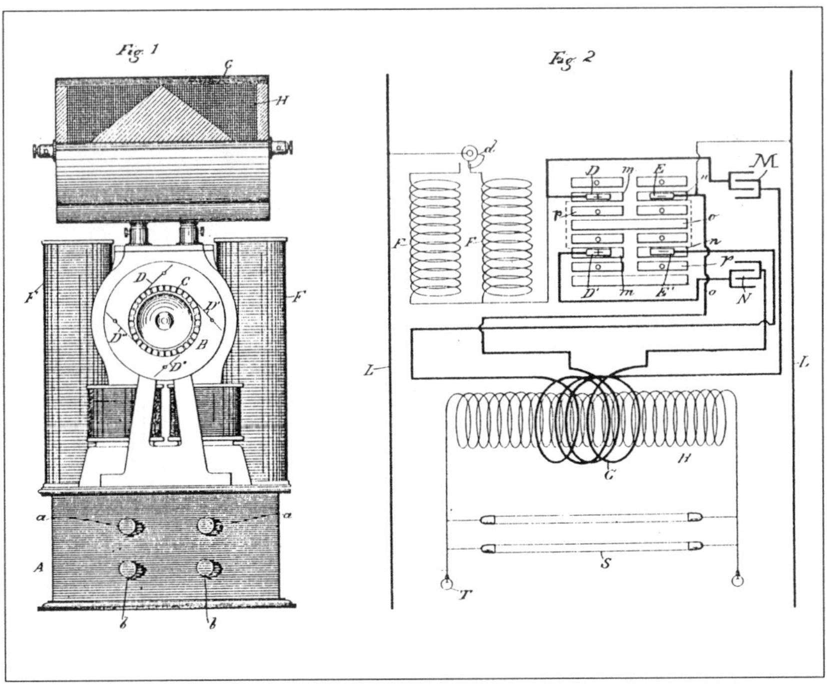

Tesla Exhibit, Working Drawing No. 1 Sheet.

earlier date. It was, as stated, an apparatus which I used in other experiments, and which I then adopted for the purposes of the investigation of the subject mentioned I have employed it very frequently during the year 1898 and also later.

Here are three sheets of drawings, the one which I have first produced [previous page] refers to the apparatus such as shown in photograph No. 8.

> The drawing just referred to by the witness is offered in evidence and marked "Tesla Exhibit, Working Drawing No. 1 Sheet."

The two remaining sheets are working drawings of the apparatus later referred to which was subsequently adopted for the purpose of my investigations and used as stated. The box supporting the motive device for the make-and-break device contained, I think, eight condensers.

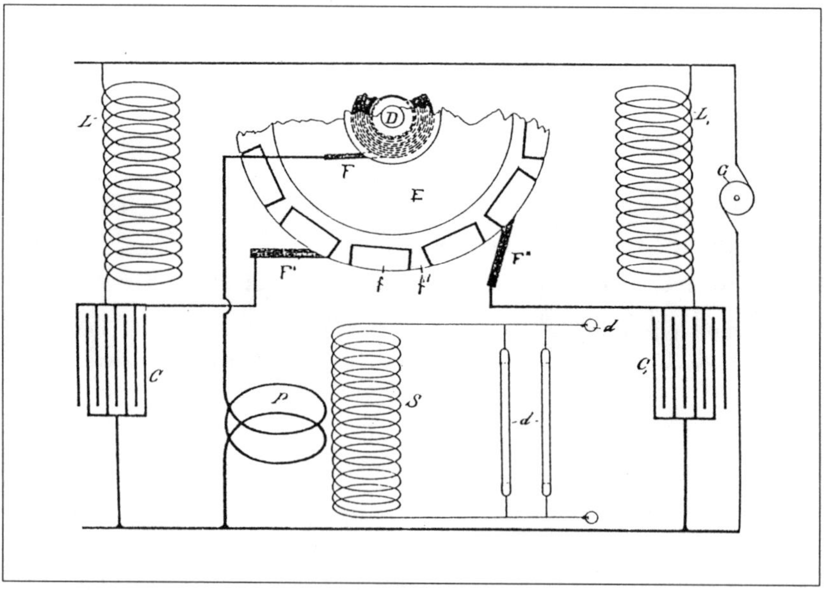

Tesla Exhibit, Working Drawing No. 2, Sheet 1.

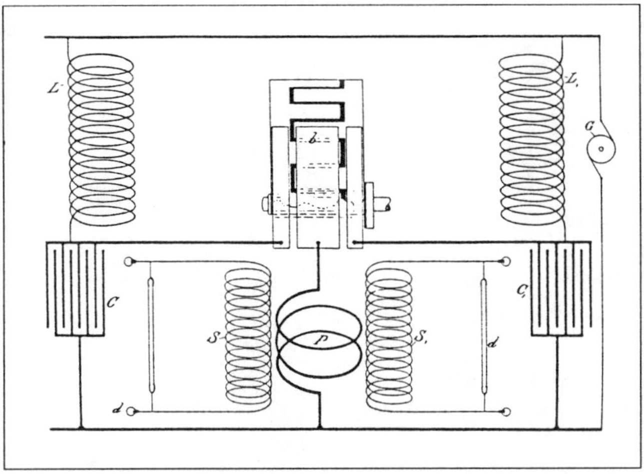

Tesla Exhibit, Working Drawing No. 2, Sheet 2.

> The drawings produced by the witness are offered in evidence and marked “Tesla Exhibit, Working Drawing No. 2, Sheets No. 1 and No. 2.”

Here [following page] is a photograph taken in Colorado and showing the general way in which I sometimes obtained two oscillations. To this mode I referred in my answer to question 39. The picture shows a large ball which formed the terminal of the transmitting circuit wound on a frame running around the laboratory, which is seen in the background, and which I used as one element of the complex transmitter. As the other element I would use a coil of which I had a considerable number, and one of which is shown in the photograph to the left of the observer, at a distance of something like 20 or 25 feet from the ball. When employing the apparatus in accordance with the invention described in my application a coil, as the one illustrated, would be connected to a terminal, such as the ball shown, with one of its ends, while the other end was grounded or, as the case may be, connected to the ground through a condenser. When the transmitter was set at work and the discharge

Tesla Exhibit, Photograph No. 9.

took place, both of the systems would vibrate, each its own period, and thus two vibrations of suitable frequency were passed into the ground, and were used to excite two receiver circuits, each tuned to one of them. This method I also used in obtaining more then two oscillations.

> The photograph produced by the witness is offered in evidence and marked "Tesla Exhibit, Photograph No. 9."

This photograph illustrates an experiment performed with the oscillatory apparatus in Colorado, and shows several transmitter coils or circuits which I have employed in practicing the invention in the ways described and also in other ways. On the right side of the observer will be seen the Westinghouse 50 kilovolt transformer, to which reference was made in the earlier part of this testimony. The frame around the room, on

Tesla Exhibit, Photograph No. 10.

which a transmitting circuit was wound, is plainly shown, particularly the top turn of the latter which was supported on glass insulators. The white streaks show the discharge passing across the hall. The photograph was made by two exposures, of course, as it would be impossible for any person to sit there while the discharge was passing.

> The photograph produced by the witness is offered in evidence and marked "Tesla Exhibit, Photograph No. 10,"

Here [following page] is a photograph of the laboratory at an advanced stage of construction illustrating the actual appearance at a time when I conducted some experiments bearing relation to the invention. I refer to a ball shown to the right, supported on an auxiliary frame, which was used in investigating the effect of the

Tesla Exhibit, Photograph No. 11.

height, and sometimes, for convenience, as an adjustable terminal for one or more circuits.

The photograph produced by the witness is offered in evidence and marked "Tesla Exhibit Photograph No. 11."

63 Q. When, with reference to your return from Colorado, were the photographs which you have produced, taken?

A. Some of them were taken before, and I think a few after, my return.

64 Q. When, with reference to the time of your return, was the apparatus which is illustrated by these photographs actually constructed and used?

A. They were all used before my return.

65 Q. You have produced no photographs or models of small signaling apparata capable of being used as receivers in practicing the invention of your application in interference. Why is this?

A. This is perfectly natural. First of all, I used a great variety of apparata and they were changed as fast as I discovered improvements. Then, again, very often I employed a variety of devices, such as relay and, telephone receivers, which I would readily adopt for my purposes. Since I embarked in a commercial enterprise, I have begun to systematically develop apparatus of my own, which has resulted from all my previous work, and which I propose to manufacture and to introduce.

66 Q. Up to the time of your return from Colorado, to whom had you described your proposed plan of applying the invention of your application to the transmission of telegraphic messages?

A. I had described it, as before stated, to my secretary, Mr. Scherff, and other assistants of mine have been made familiar with many of the apparata I have employed and have seen me perform experiments bearing on this invention. I believe that I have communicated, before my trip to Colorado, the principles involved to others, but I cannot distinctly recollect whether this is so, and I have had no opportunity to see the persons. I think, to mention one, that I have clearly explained the invention to, Captain Richmond P. Hobson before he went to China, but I am not quite sure about that. My assistant in Colorado, Mr. Lowenstein, has seen me use some of the apparata described, and I have kept him at work on a number of details relative to this invention all the time he was with me in Colorado from the beginning of June toward the end of September, 1899. In my demonstrations at my laboratory, 46 East Houston Street, New York. City, I have very frequently, in a general way, illustrated the principles involved in the invention, but I have a doubt as to whether the persons to whom I explained them understood thoroughly, though it is quite possible that they did. I have convinced myself more than once that I am a very poor demonstrator.

Adjourned to Thursday, August 7, 1902, at 11 A.M.

New York, August 7, 1902

Met pursuant to adjournment.

Present — Counsel as before.

CROSS-EXAMINATION BY MR. CHRISTY:

67 x-Q. Has anyone besides yourself any interest in the inventions and applications here in question? If so, who?

> Objected to as irrelevant, immaterial, as not cross-examination, as inquiring into matters in no way connected with the subject at issue, but into the private affairs of the witness, and he is advised that he is not under any obligation to answer the question in such a way as to disclose his private affairs in any way that he may deem prejudicial to his interest.

A. There is one obligation which I feel I ought to comply with, and that is one of courtesy, but I cannot answer this question because the support which I have obtained was given me by a great man who has requested me not to reveal him name.[6]

68 x-Q. You have said that you are at present engaged in constructing a plant, and I recollect rightly you mentioned a company in connection with that plant, is that so?

A. I do not recollect having made that statement.

69 x-Q. In answering question 5 and elsewhere in your testimony, you have used the expression "electrical oscillations." Please define briefly what this expression means and state whether or not it refers to what are properly known as Hertzian waves.

A. The term as I used it in the first lines of my answer stands to designate electrical oscillations of any

[6] Reference is to J.P. Morgan.

frequency, but I concentrated my efforts on oscillations of a much lower frequency than those which are assumed to pass through the space in straight lines and which are named "Hertzian." In most places in my testimony, when I refer to oscillations, I mean such as are able to pass through a conducting path, or a path of least resistance, however tortuous. Most of the experiments which I have described, insofar as they aim at the attainment of the results contemplated, could not be carried out with oscillations known as Hertzian waves, or such which approximate these closely in frequency. The Hertzian circuits are not susceptible to many experiments of resonance on account of the rapid waste of energy into the space.

70 x-Q. If I understand you, Hertzian waves are electrical oscillations propagated, we will assume, through the ether and having a certain frequency. Will you indicate within what limits such waves would be termed Hertzian, and in that way distinguish the Hertzian waves from those of your experiments?

A. In my own opinion the Hertzian waves are, as I stated, in the nature of radiations or rays. This is the best distinction I am able to give in answer to the question. The oscillations I have used most frequently, and which are contemplated to be used in my system of wireless transmission of energy, are transmitted by conduction and do not pass in straight lines through space, so, at least, I understand the subject without wishing to assert that my view is absolutely correct.

71 x-Q. What, then, is your theory regarding the transmission which produces the phenomena of your experiments?

A. Some of the experiments I have described are based on what is termed as electromagnetic induction, while in the experiments which are made in accordance with my system of wireless transmission of energy the propagation is effected by a process called conduction; the current, according to my theories and understanding, passing from the transmitter through the earth and reaching thus the receiving circuit, excites the same.

72 x-Q. In the experiment which you have referred to wherein your transmitter was in form of a coil extending around the walls of a room and your receiver was placed in the center of the room, was the phenomenon due to electromagnetic induction as you understand it?

A. In this particular experiment, according to my understanding, the action on the circuit in the center of the room was one of electromagnetic induction, and the currents both in the exciting and in the excited circuits were such as were capable to pass through a tortuous conducting path. In saying currents, I wish it to stand also for oscillations.

73 x-Q. Referring more particularly to these experiments which you made before leaving for Colorado, how were the transmitting and receiving instruments placed relatively to one another as a rule?

A. You mean in the experiments which were performed by the electromagnetic induction?

74 x-Q. I mean all experiments which you considered to involve directly the inventions here in question but, if it will make matters plainer, you may distinguish on the line that you indicate.

A. When performing experiments by electromagnetic induction, it is, of course, important to place the receiving circuit in a certain position with respect to the transmitting circuit. This method was referred to in my answers chiefly because it was one of the ways in which I practiced the invention. In most cases, the transmission was effected by my methods and apparata described in my patents on the transmission of energy without wires, in which it made little or no difference in what way the transmitting and receiving circuits were arranged relatively to each other. I do not mean by this that it is immaterial how the individual transmitting and receiving circuits were arranged with respect to one another.

75 x-Q. I infer that, in experimenting by electromagnetic induction, the receiving apparatus must be placed within the enclosure of the transmitting apparatus; am I correct, and, if I am mistaken, please explain this more fully?

A. No, sir; it is not necessary that the receiving circuit or circuits should be enclosed or surrounded by the transmitting or energizing circuit or circuits. But the inductive action is much stronger, because, as it is theoretically explained, a much greater number of lines of force link the circuits when they are thus arranged than when the receiving circuit or circuits are placed outside of the loop formed by the transmitting circuit or circuits. To give an idea, a powerfully excited loop is fully capable of producing an effect at a great distance, possibly of several hundred miles, and, in fact, some scientific men have advocated this method of transmission, and some have given practical data and formulae enabling to determine the effect at a distance, produced by a transmitting circuit. Personally, I have found that my system of energy transmission is superior in this respect to such an extent that I do not think it possible that those, who have proposed to practically apply the method of electromagnetic induction, although they are recognized and great authorities, will ever develop anything capable of competing with my system. But this remains to be seen.

76 x-Q. Over what distance did you in fact transmit energy by electromagnetic induction before you went to Colorado?

A. This I cannot remember and could not state. I only can say that I frequently passed through the cable around my room energy at a rate of from 2,000 to 3,000 horsepower and that I actuated devices on various floors of my building by the loop so excited. I frequently calculated to what distance such a loop would carry; that is, at what distance it would affect a receiving circuit or device requiring a fixed amount of energy, and as nearly as I can recollect now I found that it would be capable to produce a perceptible effect at considerable distance. In making my calculations I relied on data and formulae furnished by a number of scientific men who have investigated this subject in detail.

77 x-Q. Does this transmission by electromagnetic induction and the apparatus used for that purpose involve the inventions here in question, as you understand them?

> It is for the witness to say what he did, and it is for the Commissioner of Patents to determine whether that was within the invention as defined by the claims.
>
> Question objected to as irrelevant, immaterial and incompetent.

A. As I have stated in my testimony relative to this subject, I have operated devices in this manner, or by this mode of transmission, using two oscillations and two tuned receiving circuits, the conjoint action of which the operation of the receiver was due. I am absolutely ignorant as to whether the authorities would consider this method comprised within the scope of the application I have filed in seeking protection for my invention.

Personally, I would by very much disappointed if I could not protect my invention in all its valuable applications irrespective of the manner in which the effects are transmitted.

78 x-Q. In the experiments made before you went to Colorado, in which the transmission was accomplished by what you have termed conduction, over what distance did you actually operate?

A. I believe I have answered this question in part in the earlier portion of my testimony already given. Some time in 1895 and probably in 1896 also, I made experiments between my laboratory and various places in the city, which I think might have been something like 5 miles away. It was a time when I saw that I could make a fair success with my system, introducing it on a small scale. By this system I do not mean the invention here involved, but my system of transmission with one single tuned circuit on the transmitting and receiving end. As I stated before, I made discoveries which enable me to reduce the determinations of the effects produced at a distance to simple formulae or rules of electrodynamics. Having found these laws to be rigorously true in certain respects, further trials of this kind became unnecessary, and the dominating idea became to perfect a powerful transmitter, and a little later there arose the necessity which lead me to the clear conception and practice of the present invention.

79 x-Q. Over what distance did you, before going to Colorado, actually impart impulses in which the receiving apparatus was made to act by the co-operative effect of a plurality of impulses of different periodicities?

A. I could not tell that exactly, but I remember some experiments I made operating between buildings in the city which may have been possibly a mile or two away. To this I attached obviously no importance for the reasons I have already given. The laws of propagation were the same whether I operated with one or many circuits.

80 x-Q. Can you specify an instance such as you refer to in your preceding answer?

A. This has been some years ago, and I can only in a general way. I remember that I went to various buildings in the neighborhood of my laboratory in Houston Street, but these were so many that I cannot tell now any particular instance. Perhaps in thinking over it and asking my secretary, Mr. Scherff, and perhaps some other assistants, I might get a clue and specify some particular instance.

81 x-Q. In answering question 10, you give as one reason for going to Colorado that you "wanted to be free of the disturbing influences in the city," and that you recognized a necessity for carrying the invention in question to a further state of perfection. Did this further experimentation which you desired concern matters effecting the practicability, I mean the commercial practicability of the inventions in question?

A. I do not think that it is novel to those who are acquainted with delicate electrical instruments, such as tuned circuits, that they are very frequently disturbed in the city, such as New York, through which there are continuous leaking currents from power stations, and which is subjected to numerous other disturbances capable of affecting the working of instruments. The perfection of tuning, considered as an art, can be carried forward indefinitely, at least in theory, and no one can say, however skillful he be, that he has reached the ultimate perfection which it is possible at attain. Tuning is my specialty, and it is very likely that I might have done better than many others, yet I felt that I ought to still further improve in the knowledge of the same and in the skill of applying it. The system, as I practiced it, before

I went to Colorado, was fully capable of being commercially applied, but, of course, in order to realize what I had proposed myself — the transmission of signals across the seas, and to any terrestrial distance — I had of necessity, for this reason alone, to leave the city. To begin with, I could not have operated in the city a transmitter producing disturbances of such immense power as I contemplated to obtain, and then again the measurements and other observations which I had to make in gradually perfecting the apparatus could not have been made with that degree of precision of which they were capable in such a locality as the one which I had selected. These were, however, only a few of a great many reasons which prompted me to go there.

82 x-Q. Before going to Colorado, did you, by using the apparatus and method here in question, actually transmit intelligible communication as distinguished from unconnected impulses?

A. Yes, sir, I did. I sometimes would send simple signals, merely to convince myself that the instruments operated well; but occasionally I would transmit the signals in a certain order, which was previously agree upon, and which conveyed a meaning. At times also I would use some telegraphic instrument which recorded them.

83 x-Q. Were any of the photographs which you have produced made before you went to Colorado?

A. Yes, sir; I believe I have answered this question already. Some were made after my return. To quote one, that showing the general view of my laboratory which is marked "Tesla Exhibit, Photograph No. 5."

84 x-Q. Does any one of these photographs show all of the parts of the apparatus here in question, and, if so, which?

A. I think that in one of the photographs, the last I mentioned, there may be seen the essential elements of the transmitting as well as of the receiving apparatus. By essential elements I mean the transmitting and receiving circuits. Of course, given the transmitting and receiving circuits, anyone skilled in the art will readily adapt these for operating such instruments as he might contemplate to employ. But if this photograph shows a complete complex transmitter and a complex receiver, it is only by accident, for it was not intended, and I do not think that I ever intended to take a photograph of such a complete apparatus. Referring to the photograph, "Tesla Exhibit, Photograph No. 5," there will be seen, in the background, the spiral coil wound on a frame 8 by 8 feet which I have repeatedly mentioned. To the left of the observer there is a wooden frame which contains two turns of primary cable capable of being connected in series or multiple arc. Some condensers are seen farther to the left near the iron column. This is the apparatus with which, as I have already stated, I used to produce two oscillations when transmitting the energy by conduction. Very far back is shown a coil which, I think, I recognize as one of those used in some experiments to which I have referred. Now, on the molding at the left, there is the cable laid around the room, through which I passed two oscillations when operating by electromagnetic induction, and quite in front there is the apparatus to which I have already repeatedly referred, having two apparata with the condensers forming, as will be seen, a complete system comprising a complex transmitter and a complex receiver, but the receiving devices proper are not shown in the photograph. These were, as I have stated, very frequently varied.

85 x-Q. Does the photograph you have just referred to show means for imparting to the transmitter two or more sets of impulses of different periodicities?

A. It shows, though indistinctly, some condensers and other apparatus. Of course a clear illustration of all parts in a photograph would be almost impossible, but a number of means of producing oscillations of different periods simultaneously or successively were, as I have pointed out in the text of my application, well known to experts at that time when my earliest experiments, here recorded, were performed.

86 x-Q. When was this photograph made, and where and at whose direction?

A. I think that I requested Mr. Alley, of Tonnelé & Co., artist photographers, to take this picture, and the object was to see how the photograph will turn out when taken by my improved light, "artificial daylight," as I have called it, with which my laboratory was lighted. This shows my laboratory, at 46 East Houston Street, New York City. The exact date of the photograph I cannot give, though I may be able to find it out. I think it was taken a year and a half ago, but I speak from memory. The majority of the apparata — I might almost say all — which are visible, were constructed before my going to Colorado.

87 x-Q. You have referred to Figures 9 and 10 of the *Century* article of April 1895.[7] Did you say that these two figures show two receiving coils adapted to receive two sets of impulses of different periodicities?

A. There were two circuits in the apparatus shown in these figures. In Figure 9 a loop of wire will be seen, while below it there is a coil of comparatively many turns. These receiving circuits usually have been adopted by me for practicing the invention in the manner described, if I would have had my present knowledge, or even the knowledge which I have had at the time when I began to practice it. As stated before, this apparatus was destroyed by fire, and another apparatus, already described, was constructed in my laboratory at Houston

[7] Illustrated at p. 16 in response to question 23.

Street, I think sometime in 1895. This was the apparatus – to be more explicit – which I have used in actuating by electromagnetic induction, a device through the conjoint action of two tuned circuits.

88 x-Q. Please state whether or not Figure 11 [page following] illustrates the lamp receiver which you have already referred to in your testimony? If it does not, please explain wherein it differs.

A. No, sir, this lamp was lighted by the action of one single circuit, which is below under the cloth The circuit which energized the loop held in the hand of the experimenter was, I believe, covered with black cloth, so as to get a better picture. The photograph was, I think, also taken by Mr. Alley.

89 x-Q. In your answer to Question 16, you say of the large coil to which you have frequently referred, "This coil I excited by a primary so proportioned that when the primary was closed by the make-and-break disk which discharged the condensers, the oscillations in the secondary were quickened much above the rate which the secondary or spiral vibrated when the primary was opened." Please explain this statement more fully?

A. In endeavoring to perfect my coil, and by this I refer to a form of oscillatory apparatus which I have originated, I discovered that at times I would get two oscillations, and I soon found the cause. The oscillation of the spiral coil in this case was obviously dependent on its coefficient of self-induction and was the slower the greater the inductance which it possessed when in action. Now the induction was reduced when the primary or exciting circuit was closed through mutual induction of the circuits. This influence is readily determinable in some forms of circuits and can be calculated in advance with tolerable precision. When I stated here that the primary was proportioned so and so, I meant that the

NOTE: The illustration on this page and accompanying commentary are not a part of the formal transcript of interference depositions.

The cross-examine question 88 by Mr. Christy refers to Figure 11 of the article, "Tesla Oscillator and Other Inventions," by T.C. Martin in the April 1895 issue of the *Century* at page 928. This figure is reproduced here, together with the original accompanying caption, for purpose of reference.

Fig. 11. Experiment illustrating the lighting of an incandescent lamp in free space by induction from coil below, energized by distant circuit around the room. The loop of wire carrying the lamp is held by Mr. Marion Crawford. (From flash-light photograph.)

primary effect was so adjusted that when the primary was closed by the make-and-break disk, the self-induction of the spiral coil was reduced to a certain value which I desired to obtain. This proportioning was done generally by linking the primary circuit with the spiral coil more or less, thus regulating by mutual induction the self-induction of the coil so that I obtained two oscillations which followed each other in extremely rapid succession. I wish to add that this was not the only way I graduated the effect to obtain the desired result, for I adopted frequently other expedients when they were convenient. For instance, I would vary the self-induction of the primary by adding to it a few turns of a regulating self-induction coil, which I had on hand. It consisted of about 24 turns of stout wire wound on a drum and permitted me to vary the self-induction from a few to about 100,000 centimeters.[8] At times, I would modify the self-induction of the primary or secondary in other ways, many of which would, of course, suggest themselves to any expert who would have concentrated himself with as much energy as I did on becoming proficient in the manipulation of tuned circuits.

90 x-Q. Was this transmitter coil, which you have so frequently mentioned, built for the purposes of these inventions?

A. No, sir; it was not. It was simply adopted to practice the invention.

91 x-Q. Will you please describe more exactly than you have the receiving part of the apparatus which you used before going to Colorado? If you used more than one form, describe that form which you found most sensitive in your experiments?

A. As I explained before, I have used a great many forms of sensitive devices and receiving instruments.

[8] 100 microhenrys.

Many of these did not involve a novel principle original with me, while others did; at least I think so. One of the common forms of receiving apparatus used by me I may outline as follows:

There were two clockworks which rotated arbors, each at a certain speed determined by previous experiment as most suitable and ensuring the best action of the devices I employed. On each of the arbors, driven by the clockworks, I would mount a small cylinder comprising two terminals between which there was a space, and in this space I would insert a certain quantity of small chips of metal made by a tool, so as to be as nearly as possible of the same size and shape. These chips I would place in a solution of acid, or, in general, in a solution of some kind, which would produce a layer of oxide, and the best thickness of this layer I would determine by careful experiments. Each of these small cylinders was thus nothing more than a self-restorative dielectric film, and inasmuch as these devices may suggest similar devices used in so-called Hertzian or spark telegraphy, I wish to explicitly point out that when thus constructed I found them to respond with great precision, whereas, when in place of the oxidized chips some powder was employed, the action was not nearly as precise. Now, these two devices I would connect a great many ways, each of them to one of the tuned circuits, and I would strain the minute films of dielectric in each of them by batteries, the strength of which was carefully graduated, so that each dielectric film was at the point of breaking down, and would give way just as soon as the oscillation in each of the circuits, which were tuned to respond to the several oscillations of the transmitter, would reach a certain intensity. The circuits and devices included in them were connected and associated in a great many ways, which I constantly varied in my tests and experiments for the purposes of advancing in the perfection along the lines which are here under consideration. A very common way was to include in each of the battery circuits, containing a sensitive device before mentioned, an electromagnetic relay of ordinary or special construction and then complete another circuit by two such relays, each included in one of the battery circuits. In fact, I believe that this very plan was illustrated in my application, and, as I stated therein, it was the simplest and, as I thought then, the best for carrying out the invention. At other times I would associate with the two or more tuned circuits of the receiving apparatus secondaries, which I would connect in many ways, and which, at times, I would also tune. I would use in such and similar experiments, besides electromagnetic relays, telephonic receivers, microphonic contacts controlled by the currents passed through the circuits, thermoelectric elements, small mass resistances arranged in a manner similar to that adopted by Langley in his beautiful invention of the bolometer or otherwise, or I would use spark gaps controlled in many ways, or I would utilize electromagnetic repulsions or attractions, thermic effects, and besides these devices, which did not involve any great creative faculty on my part, I also employed others which I myself originated. In drawing my application, I recollect that I selected of many the devices I have described therein partially because, on the whole, they were the simplest and easiest to operate successfully by an ordinary expert who would practice my invention. I had many others which would have been more sensitive and better suitable in some instances but would have involved special experiences in certain directions which was not generally possessed.

92 x-Q. The facts stated in your last answer all occurred before you went to Colorado, did they?

A. I do not say that. I was speaking now irrespective of my leaving for Colorado.

93 x-Q. My preceding question was directed to the time before you went to Colorado. I wish you would modify your answer so to answer that question specifically?

A. I now better understand the question. I was under the impression that I was asked to describe more clearly some apparatus I had already described, and to make a statement of a general character. It would be difficult for me to say which of the many devices I employed before my trip to Colorado was in every respect the best, but the devices described I have found as the simplest and most suitable in carrying out the invention for the reasons stated. These I used, prior to my leaving for Colorado, very frequently, and sometimes my demonstrations with them were witnessed by others.

94 x-Q. In the experiments which you performed before you went to Colorado, did you employ elevated terminals for the transmitting and receiving apparatus?

A. In those experiments in which I used the mode of transmission described in my patents on the transmission of energy without wires by conduction, I used elevated terminals.

95 x-Q. You have spoken of a device for receiving impulses which I understand to have been an electric lamp whose terminals were connected with two receiving circuits tuned to the two transmitting circuits. Is my understanding right?

A. I remember that I often performed experiments in which I have used what I have possibly called a lamp, but which may have been simply a resistance of small mass, such that when the currents through both of the receiving circuits were of a certain intensity it would be heated and modified, and the effect was so graduated that when only one of the receiving circuits was energized by the transmitter the resistance was heated in a much smaller degree, and thus another device, dependent on the operation of two circuits, was not actuated when only one of the circuits was energized. I think I ought to explain what led me to make these experiments. I remember now that on some occasions I wanted to find out how far I could utilize, for certain purposes, such apparatus in bringing about the operation of a number of devices.

96 x-Q. I still fail to see why the oscillation of one of the circuits alone would not cause the operation of this detector. Will you make that point plain?

A. Obviously, in this experiment, if one of the oscillations would have been intense enough, the resistance would have been heated as much, and possibly more, than when both circuits were energized to a certain degree. I have already said, in the early part of this testimony, where, I cannot now recollect, that I sometimes associated two tuned circuits so that a device was operated by their conjoint action when the conditions under which I performed the experiment were such that the device would not operated when only one of the circuits was energized. I looked in such experiments to realize an additional feature of some value in signaling and such uses of tuned circuits, and this was to operate, through the same medium or channel, a great number of instruments by graduating the effects.

I will illustrate this by practical example with which I have been sometimes confronted with my demonstrations in my laboratory in Houston Street. When I would show various phenomena as I did on several occasions by tuned circuits, it would occur, as I have already stated, that the devices included in two or more tuned circuits would respond and would be visibly effected by one oscillation because the exciting oscillation through the primary cable around the room was so powerful that, if any of the excited circuits was of a period not far from that of the exciting circuit, there would be sufficient energy conveyed to the excited circuit to operate a device. Now, as I have already remarked, by adopting two oscillations through the exciting circuit, I was able to reduce, or to entirely eliminate, this defect. This will probably make my previous remarks clear.

97 x-Q. Did you successfully use this detector in your experiments before you went to Colorado?

A. I cannot recollect now that this kind of detector was successfully employed by me. I only stated in answer to your question that such a detector was employed by me in the way described prior to my departure for Colorado. It did what I reasonably expected it to do, but of to the degree of success with which it might have been employed, I cannot speak.

98 x-Q. In your notes, which you have produced under date of June 11, 1899, I read the following:

"The following method and apparatus for detecting feeble disturbances transmitted through a medium seem to be particularly adapted for telegraphy. The idea was followed up in New York, but the results were not satisfactory."

I gather from what follows that this refers to the form of detectors which we are now considering. Is that so?

A. As nearly as I can now recollect, when I wrote this I contemplated to investigate, further, resistances of small mass which would diminish when the current of the receiving circuit was passed through them. I distinguish between devices of small mass, the resistance of which is reduced when the current is passed through them, and between those in which there is an increase of resistance. I remember — I do not know exactly when at the present time — that I would also combine two devices, one of which would increase in resistance and one of which would diminish, and sometimes I would connect such devices in a bridge, resembling the Wheatstone arrangement, and this I would effect some kind of instrument. I think that, in this particular sheet I must have meant a device which, when the current was passed through it, diminished in resistance, and I probably meant here that the detector as used, insofar as its change in resistance is concerned, was not successful in my earlier experiments before I left for Colorado. I think I began to conceive certain ways, new to me at that time probably, in which a device on this principle might be made successful in operation. This was one of the sheets penned as usual for the purposes which I have already specified in my reference to the sheet of June 27 first produced. I do not believe that I meant in this note exactly the same device which we are now considering.

99 x-Q. Are these notes, one sheet of which is in evidence and marked "Tesla Exhibit Colorado Sketch," your original notes, or are they transcribed from the original?

A. Most of these notes were the first written records as I made them on the dates given above on each sheet, but sometimes, when I would carry on experiments and would have soiled hands, for instance, I would take a pad and first draw a sketch, make a calculation or a rough description of some kind, and then I would go to the desk and put it on a sheet. It is possible that in some of these cases I might have done this; but, if so I would have recorded it in all likelihood. I do this as a rule. When a subject comes up that I have already recorded, I mark it as being an elaboration or transcription of the previous record. These notes were all written down on the dates marked on the top of the respective pages, but it is possible that, as it frequently happens, I may have mistaken the date for one day. This, however, would be without my knowledge.

100 x-Q. Will you please state in general terms the bearing which your work in Colorado had upon your development of these inventions in question?

A. I believe I have been explicit in stating that my efforts went towards producing powerful transmitters and perfecting the methods and means set forth and claimed in my applications. In these directions I have bent my efforts all the time while I was there with the object of commercially utilizing the inventions on an extended scale.

101 x-Q. In your testimony you have said that upon your departure for Colorado you explained to Mr. Scherff that you could transmit messages by the apparatus and method here in question. I do not desire to take advantage of any inaccuracy in statements which you may make, but I think that your statement is open to the criticism that you are referring to something that you might have done but did not. Will you please make any statement in this connection which you think pertinent.

A. My phraseology in answering the question was not a happy one when I stated that I could secure the privacy and send through the earth a great many messages simultaneously. This simply referred to a degree of privacy and a number of messages which, of course, can be carried as far as one desires, it being simply a question of capital. Mr. Scherff knew that I had transmitted messages, and a number of them, and that I had obtained privacy to a certain degree, which, for practical purposes, was sufficient.

102 x-Q. Wherein did the apparatus and method which you employed at Colorado Springs differ from that

which you had previously used? The question is, of course, to be understood as directed to the inventions in question alone.

A. Of course, broadly considered, the methods were the same, although in my work in Colorado I advanced in the perfection of many of the details. As to the apparatus, I also advanced, improving the old forms, and later on introducing new forms which, insofar as the transmitter is concerned, enabled me to produce more powerful and more persistent oscillations and, as regards the receiver, permitted me to obtain a greater degree of precision and also higher sensitiveness.

103 x-Q. What form of receiving device did you find most satisfactory in Colorado Springs?

A. In Colorado Springs I perfected one or two devices on novel principles which I propose to secure by patents but of which I cannot speak now. They have not been described in my applications.

104 x-Q. Did you successfully use the lamp detector?

A. I do not know what you mean by a lamp detector. It is, I suppose, a detector of small mass as I used it.

105 x-Q. I have in mind the detector which you mention in your answer to questions. 23, 24, and 25.

A. If I understand the question rightly, you mean, broadly, a detector which was a resistance heated to such a degree as to be incandescent, to which I have previously referred in my answers to questions 24 and 25. While this was a detector in a broad sense, it would not have been suitable in tests in which a more delicate device was required. Such devices have already been mentioned here, and if you mean a detector on the principle of the bolometer, I have used it with a degree of success, but not with such success as the devices of my own invention.

106 x-Q. Cannot you specify a particular detector which you employed in Colorado Springs with satisfaction and which is not something which you consider a secret matter?

A. Yes, sir; I have already described detectors which I used in Colorado with success; these were the cylinders before referred to rotate by clockworks.

107 x-Q. Does any one of the photographs which you have produced show in its entirety the apparatus set up at Colorado Springs? If so, which?

A. No, sir; I do not think there is such a photograph. There are some parts illustrated, but not a whole set of apparatus.

108 x-Q. Please specify which ones of these photographed were taken at Colorado Springs.

A. The photographs marked "Tesla Exhibit, Photographs Nos. 2, 9, and 10"

109 x-Q. Photographs Number 9 and 10 show apparatus which operated by induction, if I understand them. Is that correct?

A. No, sir. This is transmitting apparatus in which the oscillations are transmitted to the ground and pass away by conduction.

110 x-Q. These photographs show the transmitter and receiver both, do they?

A. No, sir; they only show elements which have been used by me as transmitter elements.

111 x-Q. Are they adjusted for transmission as they appear?

A. No, sir; they are not shown in action for such purpose.

112 x-Q. What was the maximum distance to which you transmitted impulses with the apparatus and methods of these inventions while you were in Colorado Springs?

A. I transmitted impulses at a very great distance, but if you mean the distance at which I set up a transmitting and receiving plant and so operated them, I had no object of making experiments at any considerable distance. I may give you an explanation of this. Certain discoveries enable me, as I have repeatedly remarked, to calculate exactly the effect which was produced by my transmitter at great distances, and I adopted an experimental method of ascertaining the effects also by measurement. This I did by comparison with lightning discharges which occurred almost every day and which permitted me to determine the effect of my transmitter and to ascertain experimentally the energy which it was capable of transmitting, as compared with that energy which was transmitted from a certain great distance by a lightning discharge. These I could follow up to distances of many hundreds of miles, and I could at any time tell precisely how much of a fraction of a watt I would obtain with my transmitter in a circuit situated at any point of the globe. The energy ascertained by measurement agreed exactly with that determined by calculation. In some instances, I would set up receiving instruments at a distance of a mile or so from the laboratory, reduce the power of the transmitter, and operate them. A very minute fraction of the total power developed was sufficient to actuate the devices.

113 x-Q. Did you in Colorado Springs place the transmitter and receiver a mile apart and transfer messages between them?

A. Yes, sir; I did on some occasions. I mean to say, not just a mile, but more or less. For instance, I would take, in some experiments, my receiving apparatus to a small lake which was nearby, I do not know how far from the laboratory, a mile or possibly two, and would receive the signals there.

114 x-Q. And when operating at that distance, did you employ the conjoint action of several impulses of different periodicities?

A. I did, sir, on a few occasions.

115 x-Q. Did anybody see you do that?

A. I think that others saw me perform these experiments, at least one or two persons. Of course I did not desire many to see me do my work. My assistant, Mr. Lowenstein, will perhaps recollect my taking instruments and coils which I had previously tuned outside of the laboratory and observing the action of the impulses sent from the transmitter plant according to a certain order of succession. It was very convenient for me to use, in some of these experiments, two oscillations produced in extremely rapid succession by the make-and-break disk as it closed and opened one primary which was energizing the transmitting circuit. I think, also, that in some other experiments I used to generate two oscillations, likewise in a convenient way by discharging the elevated terminal of one of the elements of the transmitter into another one. Both of these ways I have already described.

116 x-Q. In this Colorado Springs apparatus when transmitting a mile, more or less, did the transmitter consist of a coil with an elevated terminal?

A. I had either one or two elevated terminals. It is possible to operate with one terminal in the manner I have described, and it is also possible to operate with one terminal in many other ways, but it is much more difficult than to work with two terminals or more. Of course the elevated terminal was connected to one end of the transmitting coil, while the other one was grounded, either directly or through a condenser.

117 x-Q. Does the exhibit "Tesla Exhibit, Transmitter Drawing" show any portion of the apparatus of the invention now in question, except the coil which you have marked *A* on its supporting frame *F*?

A. The photograph shows only the coil marked *A*. As I have before stated, the exciting primary is behind and not visible. I do not see that the cut illustrates anything else which would be important to point out.

118 x-Q. Please state as accurately as you can the date of your return from Colorado, and give your reasons for fixing it?

A. I cannot state now with absolute exactness when I came back, but I think it was on January 11, 1900. I may be mistaken, however, and can readily establish the exact date by going with my secretary over my correspondence.

119 x-Q. What do "Tesla Exhibits Working Drawings 1 and 2" show, and when were they made and by whom?

A. The drawings marked "Tesla Exhibit, Working Drawing No.2, Sheets Number 1 and 2," are drawings of an older form of apparatus which were made, I believe, by my draughtsman, Mr. Diaz Buitrago. One of them, Sheet 1, is marked August 21, 1896, and the other September 3, 1896. This apparatus I subsequently adopted to practice the invention. The third sheet, "Tesla Exhibit, Working Drawing No. 1," is, I find, not dated, but I have reasons to believe that it is also made by Mr. Diaz; when, I cannot remember. I only know that apparatus, as illustrated here, has been constructed by me prior to 1898, and that I have used it in experiments during 1898 and at a later date. These two apparata I employed when I experimented with circuits through the medium of a wire. Both of these apparata illustrate transmitters, and I employed them to produce two and, as the case may be, more oscillations when practicing the invention.

120 x-Q. Are they specially designed for producing two or more sets of oscillations?

A. I cannot now tell what guided me to make the arrangements illustrated when I designed the apparata, but I know that I did produce a number of oscillations with them and that, when I needed them for that purpose, I readily adopted them.

Adjourned to Friday, August 8, 1902, at 10 A.M.

New York, August 8, 1902.

Met pursuant to adjournment.

Present — Counsel as before.

121 x-Q. You have mentioned some signaling which you did at Colorado Springs by co-operating a plurality of impulses or different periodicities to the distance of a mile, more or less. Did you do this with the transmitter shown in your photograph "Tesla Exhibit, Photograph No. 11"?

A. I operated not exactly with the apparatus as illustrated because at the time when the terminal *S*, shown in the photograph, was put up, I had concentrated my efforts on producing as powerful oscillations as possible, and I had already thoroughly satisfied myself by previous experiments that I could secure a nearly perfect privacy, and the experiments with this apparatus, as shown, were made chiefly to advance along these lines. The ball on the right of the observer I used very often, and then it was connected, usually, to a much smaller transmitting apparatus, as the performance of the large one was too great. With this ball as a terminal and with other terminals, I operated devices. At that time, I may add, I had also perfected novel ways which permitted me to obtain the desired result by using but one circuit, instead of two or three or more, and still obtain any number of oscillations I desired. I did use, however, apparatus which is not shown in the photograph, being contained in the building, for operating devices in the manner described in my application, and at that stage I employed a terminal in the center of the building, which used to be lifted up and down on a beam to a height of something like 35 feet, if I recollect rightly, the roof having been made removable for that purpose. The removable wings of the roof are visible in the photograph, and I think I have other photographs showing the older terminal. I have stated in my testimony before that this photograph illustrates my laboratory at a certain stage of construction.

122 x-Q. At what time during the period of six or eight months which you spent in Colorado did you perform this transmission for a mile, more or less, to which your attention has been directed?

A. I can locate the approximate date readily, remembering that my draftsman and assistant, Mr. Lowenstein, left me sometime at the end of September, 1899, and I remember that I performed some experiments while he was there, and many more when he left, during October and later.

123 x-Q. I have understood that your work in Colorado was directed in a large measure to perfecting a powerful transmitter. Did you use the powerful transmitter which you had worked out there for the purposes of these inventions? If so, with what results?

A. I developed the transmitter continuously with the object of utilizing it in the spirit of an invention. I contemplated, as I before remarked, to establish telegraphic communication across the Atlantic and Pacific Oceans, and the greater the distance the more important and the more essential, I may say, it was to secure privacy and non-interference of messages; at small distances it does not matter so much whether the operation of a device is interfered with, but when costly apparatus, such as was necessary to employ in my contemplated undertaking was used to produce effects at great distances, the value of a plant and of the whole investment would be destroyed if the messages could not be kept private. I steadily bent my efforts to this end, and as far as the transmitter is concerned I endeavored to produce as powerful oscillations as possible, and oscillations of a complex character which would permit an individualization of the messages such that it was practically impossible by any one to read them or to interfere with them in any way. I may say that with the apparatus as illustrated here, I was able at a later stage to develop, if I desired, as many as one hundred different oscillations which I could all combine, if necessary, to bring about the operation of a receiver of a correspondingly complex character. This statement will, I think, be understood from the perusal of my application, but the methods of producing these oscillations by one single circuit have not been described or hinted at in my application.

124 x-Q. My inquiry is directed to actual facts, not to conclusions to which you, in your mind, arrived. Will you please confine your attention to your work at Colorado Springs and say whether, while you were there and after you had constructed a powerful transmitter, you actually sent from that transmitter impulses of different periodicities and actually received those diverse impulses on a receiver such as these inventions involve, and produced in that receiver intelligible responses?

A. I frequently used to operate with the transmitter a number of circuits, combining them as described in my application to actuate a receiver in a distant part of the building. On such occasions I intended merely to perfect the oscillations and details of the system so as to secure a greater privacy and to produce a system which would enable me to send a much greater number of messages at the same time. I do not recollect that I went outside to a distance of a mile or so and that I operated, by the improved powerful transmitter with the high terminal surmounting the shop, devices, or received messages in accordance with my method. As I stated before, at that time I had so far advanced that I felt perfectly sure of success to a high degree, and I confined myself to using this transmitter to improve it along the lines indicated and for a much greater and valuable purpose.

125 x-Q. If I have understood you correctly, the distance at which you are able to communicate according to these inventions is directly dependent upon the length of the spark gap in the transmitter; is that so?

A. No, sir. There is, of course, a dependence between spark gap and distance, inasmuch as an apparatus which is capable of giving a very long spark can charge a condenser to a high pressure, and as the energy stored in a condenser is dependent on the square of the pressure and, we may say, on the length of the spark gap, the greater the spark gap which the apparatus is capable of breaking through the greater will be the distance at which it will convey the messages, if properly used. In my system, however, as I have described it in my patents, I avoid the spark gap in the transmitting circuit because it does not permit a powerful oscillation to take place, as the energy of the oscillation is frittered down in the resistance of the arc. The distance at which in my system the messages can be conveyed by my transmitter, irrespective of the number of the oscillations of various periods it develops, is proportionate to the quantity of electricity displaced. In another way, I may say it simply bears proportion to the current which the transmitter passes into the ground.

126 x-Q. In this system, then, as you have described it, the current actually flows from the transmitter through the ground to the receiver; is that so?

A. Yes, sir; it does, in accordance with my understanding. In my Patent No. 649,621, Apparatus for Transmission of Electrical Energy, it is stated distinctly:

"It is to be noted that the phenomenon here involved in the transmission of electrical energy is one of true conduction and is not to be confounded with the phenomena of electrical radiation, etc."

RE-DIRECT EXAMINATION BY MR. PAGE:

127 Re-d. Q. I hand you copies of four patents dated November 5, 1901, and numbered 685,953, 685,954, 685,955, 685,956. Please state if these four patents were applied for by and granted to you.

A. Yes.

128 Re-d. Q. Where and by whom were the applications for these patents prepared?

A. By myself and Mr. Leonard E. Curtis in Colorado Springs, from June to August, 1899, I believe.

129 Re-d. Q. Would the note under June 11, 1899, referred to in cross-examine question 98, relate to any devices used in connection with the methods or apparatus of the patents referred to in the last two questions as well as to the inventions of the applications now in interference?

A. Yes, sir. Of course, the devices referred to in the note mentioned were suitable, insofar as their use as sensitive devices is concerned to be applied to the methods and apparata described in the patents last quoted as well as to the invention concerned in the present interference. I have used such detectors in developing both of these methods and apparata.

130 Re-d. Q. To what extent is the same true with respect to the relays, sensitive devices, and other appliances which you used in connection with the tuned circuits in practicing the invention of your application in interference?

A. The majority of relays, detectors, and such appliances were, of course, equally suitable for use with one or the other method.

131 Re-d. Q. Have you used them differently in practicing both?

A. I have, many of them.

132 Re-d. Q. Will you, for the purpose of illustrating the specific construction of these devices to which you have referred in your cross-examination, produce some specimens of such devices which you used prior to going to Colorado, or which are in all substantial respects like those which you have testified that you did so use?

A. Some instruments which I used prior to my leaving for Colorado and others which were forwarded to me while there in June, I think, of 1899, and which I have used in practicing the inventions, have been sent to me from my laboratory at Wardenclyffe, and I can produce them.

> The patents referred to in Re-d. Q. 127 are offered in evidence and marked "Tesla Exhibit, Patents, November 5, 1901,"[9]
>
> It is agreed that uncertified copies of these patents may be used with the same force and effect as the originals.
>
> Counsel for Tesla gives a notice that at the hearing of this cause reference will be made to the files of the applications for the above-named patents.
>
> The instruments produced by the witness in reply to the last question are offered in evidence and marked "Tesla Exhibit, Receiving Instruments, Nos. 1, 2, 3, 4, 5 and 6, respectively."[10]

133 Re-d. Q. Will you please describe these devices now before us, stating, as briefly as practicable, the nature of

[9] Refer to Appendix C.

[10] The Interference file carries no photographs of these instruments.

their construction and the manner in which they are used, or were used by you?

A. These instruments, "Tesla Exhibit, Nos. 1 and 2," are instruments of the most common form which I have already described in my answer to question 91. They are clockworks driving arbors on which there were mounted the sensitive devices described. I had, prior to my going to Colorado, a considerable number of apparata of this kind, and I very often used them in practicing the invention. I likewise employed them in my experiments in Colorado in practicing this invention, as well as the methods described in the four patents of November 5, 1901.

Instrument No. 3 is a modified form of device which I frequently used and of which I had several constructed, almost identical to this one.

Instrument No. 4 is another device as the Nos. 1 and 2 before mentioned.

Instrument No. 5 again is another one with a relay mounted on the base, and

Instrument No. 6 is a relay which I very often employed to close a circuit common to two or three receiver elements or circuits.

I ought to state distinctly that, as nearly as I can recollect now, all the present devices were constructed before I went to Colorado and were used in practicing the invention. There may have been springs or blinding posts mounted on them since as I have used them in Colorado. This I cannot remember.

134 Re-d. Q. Are these all the devices of this character which you have in your possession and which were made or purchased by you prior to the time when you returned from Colorado?

A. No, sir; I have a great many, and could no doubt find some more of that period. These were sent on at my request yesterday from my laboratory at Wardenclyffe. Great many of such devices have been used by me in ever varying experiments, and I do not know just at what stage of experimentation they were put aside or adapted to one or other purpose.

135 Re-d. Q. Have you, since our last session, ascertained any facts which enable you to fix with certainty the date of your return from Colorado, in January, 1900?

A. Yes, sir; I have first of all ascertained the exact date of my return to the Waldorf, which from their books is January 16th, 1900, and I have also found in a paper *Electricity*, of January 17, 1900, an article purporting to be an interview, stating that I left Denver for New York on January 13th, 1900, which agrees, according to my present recollection, with the record made at the Waldorf.

> It is agreed that the number of *Electricity*, New York, January 17, 1900, referred to by the witness, may be produced and referred to at the hearing of this cause with the same force and effect as though it had been duly offered in evidence as an exhibit in this cause. It is understood, however, that counsel for Fessenden does not admit the sufficiency of the publication in question as evidence of the fact inquired about.

RE-CROSS-EXAMINATION BY MR. CHRISTY;

136 Re-x-Q. Which of these exhibit instruments which you have produced contains the sensitive device or sensitive devices which were actually used before you went to Colorado Springs in the combination here in question?

A. Some of the instruments shown have the sensitive devices attached, as for instance, the one marked No; 1, No. 3, and No. 5. I have a number of sensitive devices here which used to be attached to one or the other of the instruments. In practicing the invention, I would take any two of these clockworks or sensitive devices and operate them by two receiving circuits, and sometimes more than two, and effect a closure of a local circuit by their combined action. As will be seen, a number of them will fit on the arbors of the instruments which are turned by the clockworks. The relay marked No. 6 I used very frequently to complete the common circuit.

137 Re-x-Q. Which, if any, of these sensitive devices of Exhibits Nos. 1, 3 and 5 was actually made and used before you went to Colorado Springs?

A. The one attached to Exhibit No. 3, and the one attached Exhibit No. 5, I remember distinctly and can identify.

138 Re-x-Q.I have not yet got a clear idea of the manner in which these sensitive devices operated, nor of the function which the clockwork performed in their operation. Will you please explain this so that one having these exhibits before him may understand it?

A. As I have already stated in describing these clockworks and sensitive devices, one form of the latter comprised two terminals, separated by a space between which small chips of metal or conducting material were inserted. Thus, dielectric films of minute thickness were formed which, when the device was included in the receiving circuit, or in another circuit in inductive relation to it, would be strained and broken down, and all the more readily as they were strained by means of a battery to the point of breaking down. The rotation of the clockwork, or, as the case may be, the reciprocal movement, would tend continuously to restore the dielectric films and so they were made self-restorative, and the receiving device was operated only so long as the receiver circuit was energized.

139 Re-x-Q. Were these sensitive devices made for any purpose other then the inventions here in question, or used in any other combination than that here in question?

A. As far as I can recollect now, they were made simply to serve the purpose of sensitive devices or detectors, and I used them in various ways and for various purposes.

140 Re-x-Q. Then, if I understand you, there is nothing peculiar about them which identifies them with these inventions and nothing else, is there?

A. The individual device exhibited I can remember and identify; but, as stated in my testimony already, I did not consider them devices as involving any particular invention on my part. So far as the devices themselves are concerned, they might be used with any receiving circuit.

141 Re-x-Q. The systems which form the subject matter of the patents produced in your redirect examination are also capable of being used selectively, are they not?

A. The patents referred to aim at the transmission of electrical energy and detection of feeble disturbances; and as to the feature of selection, I have illustrated in them devices, if I recollect rightly, which, if used intelligently, can be applied to telegraphy in a certain sense coming within this term. But not in the same sense as contemplated in my invention in interference, which I based on the use of tuned circuits.

142 Re-x-Q. Did you do any experimenting in Colorado which involved the systems of these patents?

A. Yes, sir, I did, and I frequently used in my experiments the very devices which I have here exhibited.

143 Re-x-Q. And these devices, that is, the coherers and receivers, were adapted for use and were actually used in the systems of these patents; is that correct?

A. They were used as receiving instruments or sensitive detectors simply, but not in the same sense as in the invention involved in this interference.

144 Re-x-Q. If the system now in interference was worked out before you went to Colorado, why didn't you make application for patents for it when you made applications for patents involving the other systems?

Objected to as irrelevant and immaterial.

A. I believe I can answer that, giving the reasons which are responsible for my filing the applications for the patents of November 5th first.[11] To begin with, the invention involved in this interference was very much more difficult to put in clear and precise terms, and in the transmission to great distances the methods described in the four patents named were more readily applicable than those involved in the present interference, inasmuch as the methods in the patents named enabled the detection of impulses however feeble; and so it came that although I worked on both of these methods simultaneously, the other ones were first recorded in the Patent Office. I filed the applications on the invention in interference as soon as I was able to manage it on my return to New York, as stated.

145 Re-x-Q. Is it not true of the system of these patents referred to that a synchronizing of the receiving apparatus is essential?

A. No, sir; the methods described in those patents, as I said, contemplate the transmission and utilization of

[11] That is, Tesla Exhibits Patent Nos. 685,953, 685,954, 685,955, and 685,956 offered in evidence as "Tesla Exhibit Patents, November 5, 1901, and reproduced in Appendix C.

electrical energy. I use devices for the purpose of directing the impulses and charging the condenser, thus storing the energy in such amount as necessary to cause the operation of the particular device or receiver which is used. There is no necessity that in this system there should be tuned transmitter and receiver circuits, although in connection with the methods described in some of the patents I may use also tuned circuits.

146 Re-x-Q. By synchronizing I had in mind the speed of the clockwork receiver. Is it not a part of this system, as described in those patents, that the receiver may be thus synchronized and the system thus rendered selective?

A. In the methods described, clockworks, or such synchronously-running devices are illustrated, and it is, of course, necessary that their speed should bear relation to the impulses of the transmitter, but merely for the purpose of enabling them to be accumulated in the condenser. Although by such arrangement of mechanical devices, rotated in synchronism and permitting impulses to be sent and received in certain desired orders of succession, many messages can be sent through the same channel; such system would probably not be called selective in the sense as it is used in the applications of my invention in interference.

147 Re-x-Q. Is it not selective in this sense, that a receiver speeded at a certain rate will receive the message or impulse, but if speeded at another rate it will not.

A. This might be so, but in the invention in interference receiver elements are employed which, by reason of a quality inherent in them, are selective.

148 Re-x-Q. Did you, in experimenting with the other system of these patents, employ tuned circuits as well as the synchronizing referred to?

A. I did employ tuned circuits in Patents Nos. 685,954 and 685,956.

GEORGE SCHERFF, a witness called on behalf of Tesla, having been first duly sworn, deposes as follows:

1 Q. Please state your name, age, residence and occupation.

A. My name is George Scherff; 29 years old; secretary to Mr. Nikola Tesla; residence, Wardenclyffe, Long Island.

2 Q. How long have you been associated with Mr. Tesla?

A. I have been in Mr. Tesla's employ since October, 1895.

3 Q. During that time what opportunities have you had for observing his work and experiments?

A. I have attended many lectures and demonstrations made by Mr. Tesla, and he has also made privately many explanations to me.

4 Q. You have held a confidential position and intimate relation with Mr. Tesla in his work?

A. Mr. Tesla has very frequently honored me by communicating to me his plans and explaining to me his experiments.

5 Q. Have you read Mr. Tesla's applications, which are involved in this interference, and do you understand the methods and apparatus which are made the subject of those applications?

A. I have not only read the applications, but Mr. Tesla has dictated them to me, and I have made a number of copies of them.

6 Q. When did Mr. Tesla first dictate to you these applications, or any part of them?

A. Shortly after his return from Colorado, in the beginning of 1900. He brought part of the applications with him which stated the invention clearly. I believe that part contained a clear description of the invention, it was up to where the description of the drawings begins.

7 Q. Do you know the date on which Mr. Tesla returned from Colorado, as stated in your last answer?

A. I do not remember exactly, but I have ascertained from newspapers and journals that it was about the 16th of January, 1900.

8 Q. How long after his return from Colorado did Mr. Tesla give to you or dictate to you any portion of the specification or description of the invention which is here in controversy?

A. It was during the first few weeks after his return. He then dictated a portion of the specification in addition to the part he brought with him from Colorado. Then he had to drop the work on account of an article which he wrote for the *Century Magazine*.

9 Q. I hand you now a typewritten paper (handing the witness "Tesla Exhibit Draft of Specification").[12] Will you please state, if you know, what that paper is?

A. I recognize this as one of the copies I have made. This copy was made later, I think about the time the application was filed in the Patent Office. Mr. Tesla had made corrections on the old copies and wanted a clean one for record.

10 Q. Please examine the contents of that paper and state if you recollect what portion of it was first given to you by Mr. Tesla to transcribe on the typewriter?

A. As nearly as I can remember, Mr. Tesla brought

[12] Refer to Appendix D-1 for the text of resulting Tesla U.S. Patent No. 723,188 granted Mar. 17, 1903.

with him from Colorado the text contained in the first four pages, and very little of it has been changed, perhaps only a word here and there. I remember that Mr. Tesla remarked that this application was one of the best he had ever written.

11 Q. What leads you to say that Mr. Tesla brought the text of the first pages with him from Colorado?

A. He requested me to make copies, and he intended to complete the specification, but as I stated before, he was pressed for the *Century* article and had to interrupt the work.

12 Q. I meant by my question to inquire how it was that Tesla caused to be brought the text inquired about with him from Colorado?

A. Mr. Tesla stated that he had written it in Colorado.

> It is obvious that the statements made to the witness are incompetent as proof of facts, and they are accordingly objected to by Fessenden's attorney as inadmissible.

13 Q. How soon after Mr. Tesla's return from Colorado did you first see the text of this specifications?

A. It was during the first few weeks. I cannot tell the exact date, but it was very soon after his return, for I know that he was very anxious to file the application.

14 Q. Did Mr. Tesla, to your knowledge, bring with him from Colorado in January, 1900, any other papers or sketches relating to this invention, and if so, will you state what they were?

A. He brought with him a great many notes which he had made there, and I remember that he brought them with him in a satchel the first morning after his return to the office. Then, when I remarked how well he looked, he said that he had done the work of three years during his stay in Colorado, and he called me over to his desk and showed me his notes and explained to me many of the wonderful discoveries he had made, and many things he explained to me by sketches contained in the notes. Among them, also, one which illustrates how he had perfected his system of telegraphy so as to secure perfect privacy and safety against disturbance. This was accomplished by causing a number of electric vibrations of different periods to actuate one receiving instrument.

15 Q. I hand you a number of detached sheets of descriptions and sketches. Will you please state, if you can, what these sheets are and when you first saw them?

A. I recognize these as some of the notes which Mr. Tesla brought from Colorado. I first saw them the morning after his return, when he showed them to me, turning them over leaf by leaf.

16 Q. Do you recollect when Mr. Tesla went to Colorado, or have you any means of fixing the date of his departure for that State?

A. I know he was to lecture in Chicago on May 13th, 1899, and he left two days previously; that is, on the 11th of May, 1899. The date of his lecture is also published in the *Western Electrician*. I could also ascertain by correspondence, if necessary, or by telegrams.

17 Q. Had Mr. Tesla, before his departure for Colorado in May, 1899, communicated to you in any way the invention which is described in his applications in this interference. If yes, will you please state what you recall of the circumstance?

A. Yes, sir; Mr. Tesla explained the invention to me before going to Colorado. I think it was during the few months previous.

18 Q. Had you seen him practice the invention before he went to Colorado?

A. Yes, sir, I had. Mr. Tesla, was always working with tuned circuits. He would connect these in many ways and if the experiments succeeded well, he would frequently call me to show them to me.

19 Q. At the time when you first came to Mr. Tesla did you have any knowledge of the art of transmitting electrical energy without the use of wires?

A. No, sir, none at all.

20 Q. And such information as you now have on the subject you acquired from Mr. Tesla subsequently to entering his employ?

A. Yes, sir, from Mr. Tesla and also through reading the technical journals and from text books which I occasionally read.

21 Q. Among the sheets handed you in connection with question 15 is one which has been marked as an exhibit in this case and marked "Tesla Exhibit Colorado Sketch." Please examine that sheet and state if you recall whether it was among the sketches which Mr. Tesla brought in his satchel from Colorado in January, 1900?

A. Yes, sir; I remember this sheet in particular because Mr. Tesla explained the diagrams contained therein more in detail and referred in this connection to the plans which he had perfected for signaling across the Atlantic and Pacific Oceans.

22 Q. I read you now a statement of one of the issues of this interference:

> "In a system for the transmission of electrical energy, the combination with means for producing two or more distinctive kinds of disturbances or

> impulses, of receiving circuits each tuned to respond to the waves or impulses of one kind only, and a receiving device dependent for operation upon the conjoint action of the several receiving circuits."[13]

Without reference to the purpose for which this combination of apparatus was used, will you state, as nearly as you recollect, when you first saw Mr. Tesla use such an apparatus in the manner described?

A. I think I saw Mr. Tesla first use such an apparatus during the year 1898 while he was constructing what he has called a "telautomaton." This is a small boat which would perform many operations when directed by electrical vibrations from a distance, and I frequently saw Mr. Tesla connect to it various circuits and receiving instruments and direct it in this manner.

23 Q. Please describe the general character of the apparata which he used in these experiments, and state, as nearly as you can, how he connected up and used the same?

A. I don't know the exact manner in which they were connected, but I understand that there were several circuits which were connected to sensitive devices, and so that only when the several circuits were excited could the boat operate.

24 Q. Did you understand at that time — I am referring to the time when you first saw these experiments with the boat, that it was Mr. Tesla's object to make the operation of the boat dependent on the conjoint action of one receiving circuit alone?

The question is objected to as leading.

[13] Refer to Appendix D-2, Claim No. 1, of the resulting Tesla U.S. Patent No. 725, 605 granted April 14, 1903.

A. I did not clearly understand it at that time, but Mr. Tesla explained, in a general way, that he wanted to perfect a machine which would be like a human being, so to speak, and that it should have very many characteristics so that it could be recognized or called and given orders, and that, among a great number, only one would respond to his call.

25 Q. When was the first time that any question arose between you and Mr. Tesla about filing an application for the invention, which is described in his application which is now in this interference?

A. Mr. Tesla spoke about filing an application for this invention several times shortly before going to Colorado. He said that it was a very important invention, and that he ought to have it protected before going away.

26 Q. What, if anything, did you say to him about this matter?

A. I also saw the great value of the invention and tried to induce him to put in such an application.

27 Q. I will ask you to examine "Tesla Exhibit, Transmitter Drawing" and all of the photographs which have been offered in evidence in this case, and to state in connection with each one what it represents?

A. "Tesla Exhibit, Transmitter Drawing" illustrates the transmitting coil which was used in the experiments in Mr. Tesla's laboratory at 46 East Houston Street, New York City.

"Tesla Exhibit, Photograph No. 1" illustrates a transformer used to charge the condensers which formed a part of the oscillator.

In "Tesla Exhibit, Photograph No. 2," I recognize two sets of condensers which were used in New York and then sent to Colorado.

"Tesla Exhibit, Photograph No. 3" represents the telautomaton to which I have referred.

"Tesla Exhibit, Photograph No. 4" is another telautomaton from which the cover is removed showing the circuit arrangements as they were sometimes used.

"Tesla Exhibit, Photograph No. 5" is a general view of the laboratory in New York showing the transmitter. I see here a form of receiving coil which Mr. Tesla very frequently used. This coil contains two separate circuits which both terminate on the table. To these two circuits Mr. Tesla would frequently connect different receiving instruments and operate by them a relay or other devices.

"Tesla Exhibit, Photograph No. 6" shows part of the laboratory, particularly the table to which I have just referred.

"Tesla Exhibit, Photograph No. 7" shows a small oscillator used by Mr. Tesla in his experiments.

"Tesla Exhibit, Photograph No. 8" shows another apparatus employed by Mr. Tesla in the laboratory.

"Tesla's Exhibit, Photograph Nos. 9, 10 and 11" illustrate experiments in Colorado which I have not witnessed.

28 Q. How much of the apparatus in "Tesla Exhibits, Photographs 1 to 8, inclusive," was in use in Mr. Tesla's laboratory in Houston Street before he left for Colorado in 1899?

A. All of the apparatus shown in these photographs was in use before Mr. Tesla went to Colorado.

29 Q. Here are a number of small instruments which are marked as exhibits and entitled "Tesla Exhibit, Receiving Instruments, Nos. 1 to 6, respectively." State, if

you know, what these devices are and when you first saw them, or other similar to them?

A. Some of the devices named were used by Mr. Tesla in his experiments in the wireless telegraphy before he went to Colorado, and some were constructed while he was there and sent to him.

30 Q. Please state how long Mr. Tesla, to your knowledge, has used in his experiments instruments of the kind referred to in the last question and answer?

A. I believe that I first saw him use such instruments early in 1896.

31 Q. How many clockwork devices similar to these exhibits did Mr. Tesla have in his laboratory at the time when he went to Colorado in 1899?

A. I cannot tell how many, but he always was constructing and reconstructing such devices and always had a great number of them in the laboratory.

32 Q. Please answer the same question with reference to the relay, "Tesla Exhibit No. 6"?

A. He constructed some such relays and bought a great many in the market. Ever since I can remember, he has a number of such relays in the laboratory. This particular one I recognize as a very old one.

33 Q. How many different kinds of oscillations were Mr. Tesla's transmitting devices or oscillators capable of producing? I refer now to a period previous to May, 1899.

A. I understood that the large oscillators were capable of producing two oscillations simultaneously.

> Since it appears from the answer that the witness' knowledge in this regard is not due to his own observation, the answer is objected to.

Counsel for Tesla suggests that the contrary very clearly appears.

34 Q. In experimenting with his oscillator or transmitter referred to in the last answer, how did Mr. Tesla produce the two kinds of oscillations?

A. I cannot give a technical description. The oscillations were produced by discharging two sets of condensers by means of a break disk through a coil, but I cannot describe the exact manner in which two oscillations were produced, as I am not an electrician.

35 Q. Will you please take the copy of the drawing in Mr. Tesla's application, which I now hand you, and state if you have ever seen Mr. Tesla operating apparatus such as is shown in that drawing and, if so, tell as fully as you can how you saw him performing the experiment?

A. Yes, sir; I have. In New York he used one coil for producing two vibrations. To receive these vibrations at a distance, I have seen him connect several circuits and several of theseclock-works with sensitive devices and by their conjoint action operate a receiving relay or other instrument.

36 Q. Where did you see this done?

A. In the laboratory in Houston Street.

37 Q. Where were the receiving instruments placed or located with reference to the transmitter or oscillator?

A. Usually at the opposite end of the hall from the transmitter.

38 Q. Please state, with reference to these experiments, whether they were anything unusual, or whether they were experiments which were frequently conducted.

A. Such experiments were carried on almost daily.

39 Q. And during what period of time were these experiments conducted?

A. All the time since I have been with Mr. Tesla, and he devoted particular attention to them from the summer of 1898, when he constructed his boat, up to the present time.

40 Q. What provision, if any, had Mr. Tesla in the spring of 1899 for an elevated terminal for his oscillator?

A. He had a cable leading to the roof of the building in Houston Street.

41 Q. Do you know whether Mr. Tesla, prior to going to Colorado, made any experiments with receiving instruments located outside of the laboratory? If so, tell what you know about such experiments.

A. He took out instruments a number of times to different parts of the city.

42 Q. Do you know what he used as terminals for such instruments?

A. He took with him some small balloons.

43 Q. Do you know where he took these instruments and balloons?

A. Sometimes to various buildings in the neighborhood of his laboratory in Houston Street, and in one instance I remember he took his instruments to the building of Gillis & Geoghegan, at 35 South Fifth Avenue, and another time he took them to the Hotel Gerlach in West Twenty-seventh Street.

44 Q. Mr. Tesla has described a cable which was supported in his laboratory on a molding which extended around the wall. Can you tell what cable was and for what purpose it was used?

A. This cable was connected to the oscillator and was used to set up electrical vibrations in the room.

45 Q. Will you please state, as fully as you can, all that Mr. Tesla told you about this invention in controversy before he went to Colorado?

A. He explained the invention to me clearly, and I remember that in doing so he used a mechanical analog. He explained that it worked like the combination of a safe, and said that, for instance, the safe could be opened very easily if only one number of the lock had to be turned to a certain place, but the more numbers that had to be set the more difficult it would become to open the safe; and similarly, in this system of telegraphy, the more vibrations of a different period he used to affect a receiver the more difficult it would become to disturb or read the messages.

46 Q. What, if anything, did he tell you of his plans for applying the invention to purposes of telegraphy?

A. It was in the instance just cited, when he clearly explained to me the use of this invention in telegraphy for the first time.

47 Q. Did Mr. Tesla, previous to his Colorado trip, always experiment with two or more receiving circuits for controlling the third circuit, or if not, tell what was his practice in this respect?

A. I do not think that he used a number of circuits in all his experiments, but frequently I have seen him connect and use a number of coils together.

48 Q. When Mr. Tesla used two or more receiving circuits for controlling a third in his laboratory, were these receiving circuits always grounded?

A. No, sir, they were not always grounded. Sometimes he excited them by the cable around the room.

49 Q. Were they ever grounded?

A. Yes, sir. At other times he would take them near the window and connect them to the water main or steam pipes.

50 Q. Please examine "Tesla Exhibit, Working Drawings Nos. 1 and 2," and state if you know what they are?

A. These are drawings of electrical oscillators which are still in the laboratory. I understand that they were used for producing a number of different vibrations.

51 Q. Please state now what has been done by Mr. Tesla, to your knowledge, since his return from Colorado, in the development of the inventions of his applications in interference?

A. Since his return from Colorado, Mr. Tesla has begun the construction of a commercial plant for telegraphy without wires which is now nearly completed and in which these inventions are to be employed.

52 Q. Has Mr. Tesla, to your knowledge, ever constructed any complete and special apparatus corresponding to the receiving portion of the system and comprising two or more receiving circuits and a receiving device, the operation of which is due to the conjoint action of the two receiving circuits?

A. I understand that such a receiver consists of various parts; and I have seen Mr. Tesla connect up the various devices, some of which he constructed himself and others that were bought in the market. I understand that his apparatus was always constructed so that he could adopt it to various purposes.

CROSS-EXAMINATION BY MR. CHRISTY:

53 x-Q. I wish, Mr. Scherff, that you would state in your own language what you understand the inventions which are here in question to be.

A. I understand the invention is to send from a transmitter two or more vibrations, each of a different period, and affect thereby at a distance two or more circuits tuned to the different vibrations sent out from the transmitter, and to actuate by their conjoint action a receiving instrument.

54 x-Q. Have you a distinct recollection of a particular apparatus employed by Mr. Tesla before he went to Colorado which you then knew to embody the invention as you have now stated it?

A. Yes, sir, I have. This was the table shown in "Tesla Exhibit, Photograph No. 5," and also in "Tesla Exhibit, Photograph No. 6," which Mr. Tesla frequently used.

55 x-Q. Can you say the same of any other specific apparatus?

A. I remember that he had a great number of coils wound on wooden cylinders. These he connected to the several instruments and operated the receiver by their conjoint action.

56 x-Q. Can you specify any other transmitter than the cable which you knew to be adapted to the inventions and which you saw operated according to these inventions before May, 1899?

A. Yes, sir. Mr. Tesla used also as a transmitter the flat spiral coil shown in "Tesla Exhibit, Photograph No. 1," and also in the background of "Tesla Exhibit, Photograph No. 5." When using this transmitter, all the instruments were connected to the ground.

57 x-Q. Did you, before May, 1899, actually see the coil which you have just referred to used by Mr. Tesla to transmit more than one impulse of unlike periodicities?

A. You know I am not an electrician, but when performing such experiments Mr. Tesla has sometimes called me into his laboratory when the instruments were in operation. This happened prior to May, 1899.

58 x-Q. Will you please describe as fully as you can a receiving apparatus which you saw used in Mr. Tesla's laboratory before his departure for Colorado which received these compound impulses which we have been considering?

A. There were a number of them. One of these was the table with two circuits; then there was another coil on a wooden stand wound on a large cardboard drum, also containing two circuits, and a large number of wooden cylinders wound with wire, a number of which Mr. Tesla would connect with clockworks, relays, and other instruments, but I do not know the exact manner in which they were connected.

59 x-Q. Which of these forms, if you know, was most satisfactory and which was most frequently used?

A. I believe that Mr. Tesla most frequently used the table, but which of the various devices was most satisfactory I do not know.

60 x-Q. You have, I assume, witnessed these experiments which you are speaking of?

A. Yes, sir; I have witnessed the experiments very frequently.

61 x-Q. In these experiments, what was there on top of the table?

A. In these experiments, Mr. Tesla used a number of clockworks, relays, sounders, sometimes telephone receivers, and various coils and condensers which he connected in different ways.

62 x-Q. Do you understand how these several devices were connected up, and can you describe the apparatus in this particular?

A. As I have stated, I cannot explain the exact manner in which the instruments were connected, but I know that they were connected to the two circuits which terminated on the table, and Mr. Tesla would then explain to me that the coils supported under the table were receiving two different vibrations by which the receiver was actuated.

63 x-Q. Did you personally witness the experiments in which the receiver was placed outside of the laboratory, as you have mentioned?

A. No, sir, I did not.

64 x-Q. How do you know anything about these experiments?

A. When Mr. Tesla returned, he invariably told me that the experiments had succeeded, and the assistants who aided him also told me so.

> The testimony of the witness in this particular is objected to by counsel for Fessenden as hearsay.

65 x-Q. What bearing did you understand Mr. Tesla's work in Colorado to have on these inventions?

A. I understood that one of Mr. Tesla's objects in going to Colorado was to perfect a system of wireless telegraphy which was to be absolutely secret and safe against disturbance, and that these methods or improvements thereon were to be used.

66 x-Q. You have referred to "Tesla Exhibit, Draft of Specification" and said that four pages of it you recognized as matter which Mr. Tesla dictated to you.[14] Do you know anything about the other pages?

A. I did not say that I recognized the first four pages as matter which Mr. Tesla dictated to me. I said that

[14] At question 9, page 92.

I recognized them as the text which he brought from Colorado and of which I made a number of copies soon after his return. The rest of the application he dictated to me later.

67 x-Q. Have you ever witnessed the transmission of messages in Mr. Tesla's experiments with these inventions? By messages, I mean intelligible communications and not disconnected signals?

A. No, sir; I have only witnessed the transmission of signals, such as are produced by a telegraphic sounder or relay, but I think this was only because none of the assistants understood telegraphy, and dots and dashes might have been transmitted just as well as the simple signals.

68 x-Q. I understand this telautomaton to have been a sort of toy, or, better, an educational object; that is, an object in which certain results were made visible. Was it anything more than this; that is, did it have any special purpose to perform?

A. This telautomaton was a model of an invention which Mr. Tesla has patented, and the uses are explained in the patent.

69 x-Q. Referring again to this pile of photographs, have you seen them before?

A. Yes, sir; I have seen them before.

70 x-Q. Are you quite familiar with them?

A. I was present when all these photographs were taken, except those at Colorado Springs.

71 x-Q. In answering question 33, you say that you understood that the large oscillator, to which you referred, was capable of acting in a certain manner, and in answering question 52 again you say that you understand

that a certain receiver was constructed in a certain way for a certain end. To what is your understanding in these respects due?

A. When I said I understand, I meant that my statements are in accordance with explanations made to me by Mr. Tesla of the working of his apparatus.

72 x-Q. Then, if I understand you, your answers in these particulars are based upon what Mr. Tesla told you; is that true?

A. No, sir; I mean to say that my answers are based on what I have learned from Mr. Tesla.

73 x-Q. When Mr. Tesla said that his oscillator was capable of producing two oscillations at once, did you take that statement as a fact because Mr. Tesla said so, or did you, by examining the apparatus and applying your own knowledge, satisfy yourself that was the case?

A. I never doubted Mr. Tesla's statements.

> In view of the foregoing answer, the objection taken in the direct examination is repeated – namely, that the testimony in this particular is hearsay and incompetent.

74 x-Q. Am I correct in this understanding, that Mr. Tesla in his experiments when using the coil extending around the room, shown in Photograph No. 5, did not ground the receiving circuits?

A. He did not, as far as I know.

75 x-Q. Have you ever witnessed any experiments made by Mr. Tesla in which he employed more than one transmitting circuit?

A. No, sir; I cannot remember any experiments in which he has used more than one transmitting circuit.

RE-DIRECT EXAMINATION BY MR. PAGE:

76 Re-d. Q. Whatever may have been your information in May, 1899, does your present knowledge of Mr. Tesla's apparatus enable you to say whether the transmitting apparatus which he employed in the experiments with the two receiving circuits emitted one or two or more oscillations of different periods?

A. Yes, sir; I know now that for the reception of the signals he used a number of coils, some small, only perhaps an inch in diameter, and, say, two feet long, and others up to perhaps five inches in diameter and five or six feet long. I understand that these coils must each have a different period of vibration, and that, therefore, the transmitter must have emitted more than one vibration.

RE-CROSS-EXAMINATION BY MR. CHRISTY:

77 Re-x-Q. When did you first make this inference or conclusion?

Objected to as irrelevant and immaterial.

A. I do not know when this first occurred to me, but it now seems to be quite obvious.

George Scherff.

Adjourned to Tuesday, August 12, 1902, at 10 A.M.

New York, August 12, 1902.

Met pursuant to agreement.

Present – Counsel as before.

FRITZ LOWENSTEIN, a witness called on behalf of Tesla, having been first duly sworn, deposes as follows:

1 Q. Please state your name, age, residence and occupation?

A. Fritz Lowenstein; age, 28 years; residence, Wardenclyffe,Long Island; occupation, assistant to Mr. Tesla.

2 Q. When did you first enter the employ of Mr. Tesla as an assistant?

A. I came into Mr. Tesla's employ about in April, 1899.

3 Q. Where was his laboratory at that time?

A. His laboratory was at this time in Houston Street, 46, New York City.

4 Q. For how long a time did you remain in his employ?

A. I remained in his employ until the end of September, the same year.

5 Q. When did you reenter his employ?

A. I reentered the employ of Mr. Tesla about 7th of February of this year.

6 Q. In 1899, while in the employ of Mr. Tesla, were you at work in his laboratory on Houston Street during all the period?

A. No, sir. I was in Houston Street only until the end of May and left then for Colorado, from where Mr. Tesla had requested me.

7 Q. And did you remain with him in Colorado until the latter part of September?

A. I remained with Mr. Tesla until the end of September in Colorado.

8 Q. When you went with Mr. Tesla in 1899 had you had any previous experience with, or any special knowledge of, systems for the transmission of energy such as he was at work on at that time?

A. Though I was quite instructed about general electric engineering I had been familiar with wireless telegraphy and all I know about it, or at least the largest part of it, came to me by instruction from Mr. Tesla.

9 Q. Were you personally acquainted with Mr. Tesla prior to April 1899?

A. No, sir. I first saw Mr. Tesla at a visit to him at that time.

10 Q. Please look at "Tesla Exhibit, Photographs Nos. 1 to 11" and state, if you know, what such photographs represent?

A. "Photograph No. 11" shows a photograph of Mr. Tesla's experimental station in Colorado Springs.

"Photograph No. 10" gives the interior of the Colorado station under operation.

"Photograph No. 9" the same as No. 10.

"Photograph No. 8" shows one of the oscillators which I probably saw in Mr. Tesla's laboratory at 46 East Houston Street, New York.

"Photograph No. 7" is a photograph of an instrument which I remember having seen in the laboratory in New York, but without knowing exactly what it was for.

"Photograph No. 6" shows one part of the New York laboratory and a coil in operation.

"Photograph No. 5" shows the interior of the New York laboratory containing apparatus which always were in the laboratory.

"Photograph No. 4" shows the interior of the Telautomaton.

"Photograph No. 3" shows the exterior of the Telautomaton.

"Photograph No. 2" gives a look into the interior of Mr. Tesla's Colorado station.

"Photograph No. 1" shows a transformer which Mr. Tesla used as well in his laboratory in New York as in the Colorado Springs station.

11 Q. What was the nature of Mr. Tesla's work upon which he was principally engaged at the time you were with him preceding the time when you went to Colorado Springs?

A. When I came to Mr. Tesla, he kept me working on details so as to make me familiar with his work, so I had to make drawings; and Mr. Tesla explained to me by and by more things about his system of wireless telegraphy, and after seeing that I was familiar enough to understand some details about his plans for Colorado. He also began to speak about what he proposed to do there. He most of that time kept me working on different dispositions of sensitive devices and explained to me, at the same time, that the chief feature of a practical wireless telegraph system was secrecy, immunity, and selectivity; at the same time he explained to me how two oscillations are secured from one oscillating apparatus, and he showed me one of his oscillators operating when yielding two different vibrations. Mr. Tesla so used the time before going to Colorado to prepare sensitive devices for Colorado Springs experiments, always speaking about the chief objects of his going there — namely, producing very powerful transmitting apparatus and demonstrating a complete system of practical wireless telegraphy.

12 Q. Did Mr. Tesla, prior to going to Colorado, have any apparatus for producing or transmitting two or more oscillations of different periods?

A. As much as I know now, Mr. Tesla had before going to Colorado several arrangements for producing two or more periods; but what I have seen myself at that time and understood with my knowledge of that moment was the oscillator which was shown in the "Photograph No. 5." Mr. Tesla explained to me at my drawing table the way in which this oscillator yields two vibrations, and he showed me this oscillator under operation when yielding two vibrations.

13 Q. Did Mr. Tesla produce by any apparatus which he had at Colorado two or more oscillations of different periods, and if so, please state what that apparatus was and how he used it in producing such oscillations?

A. In Colorado Springs Mr. Tesla had several means of producing two or more vibrations. Mr. Tesla had two sets of condensers which were discharged into two different primaries, this giving two different vibrations. Then Mr. Tesla used coils put on the end of his secondary, which did not have the same vibration as the secondary, so that two or three vibrations have been transmitted by this means. Besides that, Mr. Tesla used two distinct transmitting circuits, both grounded and having different vibrations.

14 Q. Did you ever operate yourself, or assist in the operation of the transmitters described in your last answer?

A. Yes, sir; I assisted in their operation and sometimes was charged to operate them alone.

15 Q. When you came back to Mr. Tesla's employ in February of the present year, had you seen Mr. Tesla, or had you had any correspondence with him with reference to his work since leaving him in the latter part of September, 1899?

A. No, sir.

16 Q. Did you know the nature of the work or experiments which he was following during the interval between September, 1899, and February, 1902, while you were away?

A. No, sir.

17 Q. What do you understand to be the nature of the invention which is here in controversy? And in connection with this question, I hand you a statement of the issues of interference, No. 21,701, as stated by the Patent Office.

A. I understand that the question here in controversy relates to using two or more different vibrations on a transmitter circuit, respectively, a complexity of electrical qualities, and to use these several vibrations or complex waves to make respond receiving apparatus, which would only respond when all the vibrations intended were received or if all the qualities given to the complex wave reached the receiving apparatus. I understand that in this invention the elements of the receiving apparatus respond to the elements of the transmitter, and that only the co-action of the responding of all these elements of the receiver apparatus make the registering apparatus register.

18 Q. When did you first see a full written or printed description of this system, and under what circumstances did you see it?

A. When I came back to Mr. Tesla's employ in February, this year. Mr. Tesla showed me a European specification on this invention, and I then told to Mr. Tesla that I don't find anything new in this invention; I meant new after my Colorado Springs knowledge.

19 Q. Do I understand from your last answer that the description was that of a system or invention which was familiar to you, or what is the impression you intended to convey?

A. I must have been familiar with this invention, otherwise I could not have had the impression that I did not find this invention new to me.

20 Q. Please describe, as nearly as you can, the nature of the receiving apparatus which you saw Mr. Tesla using, either in New York or Colorado, in his experiments with a transmitter which produced two or more oscillations of different period or character.

A. When we were working on the oscillator giving two oscillations, Mr. Tesla often charged me with handling the oscillator and would go and observe the effects made on receiving apparatus at some other place. I remember exactly that he had several devices, and sometimes even a great number of them at one time, connecting up and watching the effect of the transmitter. Though I recognize the instruments I see here (referring to "Tesla Exhibit, Receiving Instruments Nos. 1, 5, 2, 4, 6, 3,") as some of those or at least exactly like those Mr. Tesla used in these instances, there must have been others yet which he used. Much of my time was devoted to details and drawings so I couldn't give now a complete enumeration of all the instruments Mr. Tesla used; but, to repeat, the ones I see here I recognize as having been used several at a time by Mr. Tesla while I was handling the transmitter apparatus yielding two vibrations.

21 Q. State where, with respect to the transmitter, Mr. Tesla had these receiving instruments when he was experimenting with them?

A. Mr. Tesla had these receiving instruments sometimes in his room, sometimes in the room I occupied, and sometimes outside the building.

22 Q. How much, if anything, did you understand during these experiments in Colorado of the relation between the transmitter circuits and the receiving circuits which Mr. Tesla was using?

A. I understood that Mr. Tesla's system of wireless transmission of energy was based upon a discharge of a condenser through a primary, or of condensers through primaries which would induce vibrations in one or more secondary circuits which were connected either directly or through a condenser to the ground; that the currents which were set up in these secondaries had to pass no resistance of any amount to get to the ground, and through the ground by conduction to the receiving apparatus. I understood that the receiving apparatus had one circuit connected directly to the ground and a local circuit induced from this one.

23 Q. What, if anything, did you understand with regard to the tuning of these circuits?

A. When I came to Mr. Tesla I didn't understand anything at all about it, but he soon showed me the great value tuned circuits, and then I understood what tuning was.

24 Q. Do you know whether it was his custom to tune his receiving circuits?

A. Already in New York, in the spring of 1899, Mr. Tesla demonstrated to me, by operation of his New York apparatus, the physical effects of tuning; and I must say that, seeing the many different coils responding to different vibrations and showing their response by great physical effects of sparks and streamers, I was deeply impressed and astonished.

25 Q. Do you know whether Mr. Tesla used, either in New York or Colorado, sensitive receiving instruments such as the "Tesla Exhibit Receiving Instruments" in evidence in connection with tuned receiving circuits, and, if so, state what you recall with regard to such use?

A. Mr. Tesla used the apparatus I see here, or similar ones, in connection with tuned circuits, many of which we had in use in Colorado, and I cannot refrain from stating as well my utmost astonishment when Mr. Tesla showed me in Colorado the effect which these circuits have on the receiving apparatus to give me striking demonstrations of the use and beauty of these tuned circuits as my admiration of the mastership with which he handled those tuned circuits. He once connected one or more of these coils to the ground and then had me called to take sparks from the upper end of these coils when the transmitter was working at different vibrations. I got no spark, or sometimes one which couldn't be recognized by the free eye, when I touched the lower part of the ground, or when I approached my hand to the upper end of a coil which was not in tune with the transmitter; but I got a spark of say one inch, for illustration, off the upper end of the coil which had been tuned to the very vibration set up in the transmitter at that moment.

26 Q. In using a transmitter which produced two or more oscillations of different character, would the receiving circuits all respond to both oscillations?

A. According to my present knowledge, I know that each of the receiving circuits was determined to take up or pick up only one of the vibrations of the transmitter.

27 Q. Do you recall now any particular occasions in which Mr. Tesla disclosed to you his plans for operating

a single receiving instrument by the conjoint action of two or more tuned receiving circuits as you now understand that system?

A. I do not remember the very instance in which Mr. Tesla had spoken to me about using different vibrations from a transmitter to respond to different circuits of the receiving apparatus in such a way that only by their conjoint action the registering apparatus would act.

28 Q. Were Mr. Tesla's conversations with you directed to disclosing his general plans or systems which he had in view?

A. Mr. Tesla honored me indeed by speaking to me of some of his plans, and discussing them with me, though the general character of our discussions have been in regard to details about either the transmitting or the receiving apparatus.

CROSS -EXAMINATION BY MR. CHRISTY:

29 x-Q. Between April and September, 1899, were you employed as Mr. Tesla's draughtsman or as his laboratory assistant?

A. I began in New York by drawing, and continued in Colorado Springs drawing in the first time, and only gradually Mr. Tesla would discuss more and more details about the very work, so that my employ with Mr. Tesla was at first one of a draughtsman, but Mr. Tesla took me very friendly into the region of his work, gave me many private explanations as to better understand his work, and was kind enough to teach me, sometimes by special experiments, about his inventions, so I gradually went from drawing to assisting in his work. Mr. Tesla often recognized my free way of discussion, and so we discussed sometimes, or very often, either many variations on details of his plant or of his plans, but this latter case was the rarer one. Mr. Tesla soon brought me to understand the two chief reasons of his going to Colorado, and I knew then that he was to perfect an oscillator which could transmit energy in a way, and on which studies were to be made to an extent, so as to enable, after the accomplishment of these experiments, Mr. Tesla to build commercial plants. The other aim of his plans, as I understood it, was to perfect and try a system of wireless telegraphy which would be not only selective, but also immune and secret. Mr. Tesla kept me all the time that I was with him on discussions about the necessity of such a system, and frequently, in connection with these discussions, spoke about details of transmitters which would yield two vibrations, so I may recall, for instance, the moment where Mr. Tesla came to my drawing board in the laboratory in New York, 46 East Houston Street, and spoke to me about his accomplishments saying that they wouldn't be of so much use for practice if there was not a practical system to use them. At the same instance he explained to me how the oscillator in the laboratory yields two vibrations, and he then led me to the same and showed me the same in operation.

30 x-Q. During the time of your employment with Mr. Tesla, in 1899, do you know whether Mr. Tesla was engaged on any work but that which concerned the system we have now to consider?

A. I am sure that Mr. Tesla did not work exclusively on this question, but I know that he devoted a great deal of his time in Colorado to experiments in question.

31 x-Q. When did you go to Colorado?

A. I went to Colorado the end of May, 1899.

32 x-Q. Will you state again what was the object of the work in Colorado, so far as it concerned this system?

A. According to my present knowledge, Mr. Tesla wanted to try in experience the system of wireless telegraphy where the conjoint action of two or several vibrations was necessary.

33 x-Q. Before going to Colorado, to what distance did Mr. Tesla transmit electrical energy by producing and receiving two or more distinct impulses, so far as your personal observation went?

A. I do not know.

34 x-Q. Do you mean you did not witness any experiments in which this transmission was made and the impulse received, or that you don't remember how far the transmission was made?

A. As I stated before, Mr. Tesla explained to me that the one oscillator gave two oscillations at a time, and I remember having seen him working different devices in the laboratory itself by action of this oscillator, as I didn't have much knowledge at that time I didn't understand all the experiments Mr. Tesla made. That is the reason why I couldn't say that I saw myself that the apparata which had been worked were connected in the way the interference demands.

35 x-Q. Did you, while in Colorado, witness experiments in which the entire apparatus of this system was actually operated?

A. I think I did, and even many times through long periods. I saw working, and I handled myself, the big transmitter sending two vibrations through the ground by two separate secondary circuits, and I saw Mr. Tesla observing and experimenting with receiving apparata where he had several, and sometimes even a great number, of receiving apparata together. I recognize some of the exhibits as those which served for that purpose and at that instant. I did not examine the way the instruments have been connected. To give a clear idea that in these experiments the complete apparatus which involves the invention in question was used, I may state expressly that the transmitter has been exactly like the one which is shown in Figure 3 of the specification in interference.[15] We had two distinct sets of condensers discharging through two distinct primaries, through a make-and-break disk similar to that shown, with the only difference that it had more teeth, namely 25, as near as I can remember, and there have been two different secondaries, both grounded; therefore, the transmitting part of the entire apparatus involved in the specification in question was not only seen by me, but handled by me many times. As to the receiving apparatus, Mr. Tesla never charged me with experimenting with it and always did that part himself whilst letting me operate the transmitter. As I stated before, in these very moments when I operated the transmitters giving two distinct vibrations, as I say, in this very moment, I saw Mr. Tesla handling many receiving apparata. I did not follow up the connections Mr. Tesla made on the receiving apparatus, but what [other] use would have been to expressly change one disposition of the transmitter where it only gave one oscillation, to a disposition by which it gave two oscillations?

36 x-Q. I infer that you do not know, from your own observation, that when Mr. Tesla had more than one receiving instrument, such as these before us, the different instruments were tuned to the different periods of the transmitter, is that the fact?

A. That is right, as I did not watch Mr. Tesla's manipulation on the receiving apparatus. But that they have been tuned to the vibrations of the oscillator follows from the fact that I personally observed many times his tuning different coils.

[15] Refer to Appendix D-1 for Fig. 3 of the resulting Tesla U.S. Patent No. 723,188 granted Mar. 17, 1903.

37 x-Q. When Mr. Tesla was using more than one receiving apparatus in these experiments, do you know whether or not he used any other apparatus to which more than one of these apparatus were connected?

A. I do not know by my own observation, as I did not watch Mr. Tesla's manipulation of the receiving apparatus, but I know that he could not have very well made any other use of so many different apparata on his receiving stand than in having them acting jointly on another apparatus.

38 x-Q. Are these receiving devices which are before you constructed on such a scale as to be practically operated, do you know?

A. Yes, sir, because, as much as I can see now, they must be some of the very instruments Mr. Tesla used in the instance quoted by me in my last answer.

39 x-Q. In these experiments in Colorado, how far away was Mr. Tesla with his receiving apparatus from the transmitter which you operated?

A. Mr. Tesla often changed details on the receiving apparatus, and we had even employed a mechanic in that town to do these changes. Mr. Tesla would at first go with the receiving apparatus in his room and experiment there, while I was handling the oscillator. Then he would take, after certain accomplishments in the room, the whole set in portable boxes and go outside of the building, leaving me instruction for continuously switching the oscillator on and off in certain intervals. In these instances, I often ran to the door to see him for a moment, and then I saw that he went as far as, say, a thousand or two thousand feet. I could not watch him closely, as I had to stick to my switch, and so it came that sometimes when just looking out to see him for a moment I couldn't see him, and then I don't know how far he went, but by the time he came back again, being to my recollection not more than the forenoon or the afternoon, you may easily build an idea how far Mr. Tesla could have gone at the time I was standing at the switch.

40 x-Q. I refer you again to the exhibit photographs which you say are photographs of objects at Colorado Springs. These are Numbers 1, 2, 9, 10 and 11, I believe. Did you actually see the buildings and apparatus shown in these photographs?

A. I saw the transformer "Exhibit Photograph No. 1" working in that plant at Colorado Springs.

I saw the apparata shown in "Photograph No. 2" exactly as they are, except that the secondary has been wound differently before I left, and that this change has been made after my departure, the difference only being in the cylindrical shape of the secondary instead of the conical.

I saw the very same apparatus as shown in "Photographs Numbers 9 and 10" working in Colorado Springs, with the same exception that the secondary framework was conical instead of cylindrical. In "Photograph No. 11," I recognize the building of Mr. Tesla in Colorado Springs, but the two towers shown there have been added after my departure.

RE-DIRECT EXAMINATION BY MR. PAGE:

41 Re-d. Q. When you reentered Mr. Tesla's employ in February of the present year, what work did you find him engaged upon?

A. When I returned to Mr. Tesla this spring, I was astonished to hear from him about the progress he made for utilizing his inventions for a commercial plant, and after he had spoken to me about the progress he had made since I left him I was much impressed by the certainty of utilizing the marvelous inventions of Mr. Tesla for commercial undertaking and to the benefit of daily life. I wish to state expressly that ever since I knew Mr. Tesla, from the beginning of my first employ until to this day, he never stopped working hard and indefatigably on the perfection of a system of telegraphy which would stand all the commercial exigencies.

42 Re-d. Q. What practice have you had in speaking English?

A. I have spent about 11 months altogether in this country, and during this stay I learned the use of the English language as I know now.

F. Lowenstein.

Part 2

Remote Control

and

The AND Logic Gate

Remote Control

The beginnings of remote control

Experimental work before the turn of the century on remote control of moving apparatus was accomplished by connecting wires from the controller to the distant mechanism. Torpedoes were guided by means of a connecting cable with the controller. Not too many years ago, toy automobiles, boats, and small airplanes were controlled by means of connecting wires which sometimes also carried the electric power operating them. Of course, the problem with such controlling means is obvious. While not so restricting for toys, long-distance control was impractical for a torpedo which would pull behind it a cable unwinding from a large reel.

Remote-controlled devices

In England, Ernest Wilson was granted in 1897 a British patent for the wireless control of dirigible, self-propelled vessels.[1] In the same year, Wilson and Charles Evans wirelessly controlled a slowly moving craft on the Thames River.[2] Three years later they obtained a United States patent on the method.[3] No photographs or clear descriptions of the craft are extant, and it is presumed to have been little more than a crude raft.

[1] Miessner, Benjamin F., *Radiodynamics: The Wireless Control of Torpedoes and Other Mechanisms*, Van Nostrand, N.Y., 1916, p. 83. (Only 1,100 copies of this now classic work were printed.)

[2] Hammond, John H., and Ellison S. Purington, "A History of Some Foundations of Modern Radio-Electronic Technology," *Proceedings of the IRE*, Sept., 1957 (Vol. 45, No. 9), pp. 1191-1208, ref. at p. 1192; ed. comment by Donald G. Fink, p. 1189.

[3] U.S. Patent No. 663,400 of Dec. 4, 1900, "Method of Controlling Mechanism by Means of Electric or Electromagnetic Waves of High Frequency," application filed July 28, 1898.

Tesla's wireless-controlled boats

Independently and nearly simultaneously, Tesla obtained a patent in 1898 for the wireless control of vessels and vehicles.[4] He exhibited a totally enclosed boat on a pond in Madison Square Garden that same year.

In the 1890s, Tesla was benefited by financial resources from partnership investors at a level unavailable to any other independent inventor. Equated with today's values, these resources would equal several million dollars. Such funding enabled him to employ various craftsman to design and construct a large inventory of apparatus and machinery.

Tesla's experimental and patent model apparatus were of the highest quality—all appropriate for exhibit purposes. A practical reason, aside from Tesla's penchant for luxuriant effects, was the necessity to impress needed investors and the news-media, and to provide models for patent applications and interferences. Several photographs of his laboratories show exhibit cases containing apparatus which demonstrated key steps in the progress of his work. It may therefore be expected as part of Tesla's manner of operation that he was able to provide the necessary exhibits whenever necessary.

The following illustration is a photograph[5] taken in Tesla's

[4] U.S. Patent No. 613,809 of Nov. 8, 1898, "Method and Apparatus for Controlling Mechanism of Moving Vessel or Vehicles," application filed July 1, 1898. This patent is reproduced as Appendix A.

[5] This photograph is similar to one taken from the *Century* magazine shown on p. 16 of the Interference depositions text. With a name change from *Scribner's* in 1884, the *Century* "had become recognized as the best of all periodicals. . . . [Theodore] De Vine, publisher, enlisted the aid of paper mills in the production of textured papers. . . . During the 1880s also came the introduction of the halftone plate, which hastened the obsolescence of the wood cut, and here De Vine was at the forefront of the initial efforts to print the new plates." It was in the *Century* that the public first saw extraordinary 150-line screened photographs in its Apr. 1895 issue of experiments performed in Tesla's laboratory on South Fifth Ave., N.Y., and in its June 1900 issue of amazing experiments conducted at his experimental station in Colorado Springs. (Quoted remarks from Alexander S. Lawson's "Century—Part I," *Printing Impressions*, Oct. 1981, pp. 42, 43)

laboratory on South Fifth Avenue, New York, in 1895, shortly before the destructive fire. In the foreground is a "tuning table," on casters, that could be moved throughout the experimental area.

An experiment in Tesla's laboratory showing an effusion of sparks between condenser plates mounted on a coil table. The coil in the center of a large room is unconnected with the the energizing circuit. At the rear, above a backdrop curtain, an exhibit case is seen.

In addition to the wireless-controlled boat demonstrated at Madison Square Garden in 1898,[6] Tesla constructed another, larger boat illustrated below.

[6] Interference transcript, p. 39 photograph.

The boat shown here, demonstrated at the Commercial Club in Chicago, and the smaller boat shown at Madison Square Garden were constructed in the 1897-98 period. This larger boat was loaded with sufficient ballast to allow little of its profile to appear above the water.

A top-view interior photograph of the larger boat[7] is quite similar to a patent drawing showing the interior arrangement of the smaller apparatus.[8] These are reproduced together for comparison.

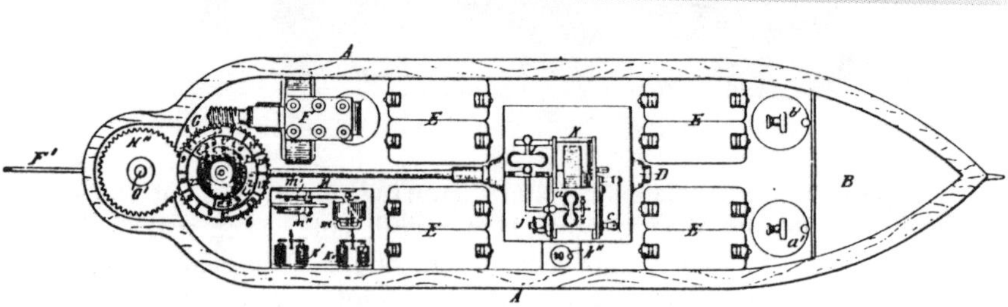

The corresponding arrangement of apparatus within the large boat is readily seen when compared with that of the smaller boat shown in Tesla's remote control patent.

[7] Interference transcript, p. 40 photograph.

[8] Appendix A, Fig. 1.

Researchers have been perplexed in not finding any newspaper accounts of the Madison Square Garden demonstration by Tesla, which followed the demonstrations there in May, 1898, of other experimenters' ideas for exploding torpedoes to protect U.S. Naval ships. The Tesla demonstration was a private showing for investors. The mode of financing his experimental and developmental work involved impressing individuals of wealth in becoming shareholders in his companies, contrasting with other scientists and inventors who were captives of corporate enterprise and academic institutions. John Oliver Ashton, friend of inventors and investors, recalls the Madison Square Garden demonstration:

> From the 1890s when I first met Tesla at the Garden in New York and he gave dad, Earle, and others, such as Morgan, Vanderbilt, Westinghouse, Hammond, Cheever, Manson, and Candee a private showing of his radio-controlled torpedo boat, I was thrilled with the idea of radio. . . . Earle was assistant to Tesla and also a financial backer from 1898-1910. I knew Tesla rather well. As the son of a wealthy manufacturer and in love with wireless and willing and able to travel anywhere to find the thrill in it, I did have a great time. Earle and I were Tesla's men by adoption. Candee and Cheever, tycoons of Okonite Wire, were quite European-minded and Tesla as a typical European gentleman-scientist appealed to them. . . . My grandfather was the original backer of Bell in Salem, but Bell made a flop of the Harmonic telegraph and felt bad because Elisha Grey beat him. That "mistake" made Bell.
>
> Those who worked for Nick were rarely taken into his confidence. His sketches were always in parts. His records were never complete. He feared pirating. . . . Louis Pacant [sic], my old chum, used to visit Tesla. As a quiet-mannered and discrete consulting engineer, he [Lewis] soon found Nick to be a companion of sorts. They would chat at Nick's penthouse but Louis could get [only] so far. Nick had the peculiar reserve and suspicion that made him a unique figure. There was no other as close as Louis.[9]

[9] From private communications to L. Anderson dated July 17 and Sept. 22, 1953, and a private communication to H. Goldman dated Oct. 5, 1954. John Ashton's father, CEO of Okonite Wire and Cable Co., gave financial support in

Tesla made a public announcement of his radio-controlled boat the day the U.S. Patent was granted to him, November 8, 1898, in an interview article entitled "Tesla's New War Wonder," appearing in the New York *Sun*. Other news articles immediately followed with headlines declaring"

"A New Marvel"[10]

"My New Submarine Destroyer"[11]

"Tesla's Peace Manifesto"[12]

"Plans to Dispense with Artillery of the Present Type"[13]

"Genius of Destruction"[14]

The news was startling. Radio was something on the forefront. Only those experimenting with wireless owned apparatus. There were no manufactured radio sets as such. And, with radio as yet unknown to most persons on the street, here was Tesla demonstrating a means of remote-control destruction by radio waves!

The news had quickly reached Europe via cable. Tesla received a letter from a dear friend who was visiting in Vienna:

> *Nov. 17/98*
>
> *Dear Mr. Tesla —*
>
> *Have you Austrian & English patents on that destructive terror which you have been inventing ? — & if so, won't you set a price upon them & commission me to sell them? I know cabinet ministers of both countries — & of Germany, too; likewise William II.*

(continued) behalf of Okonite's wealthy owners to inventors such as Tesla. Earle Lewis Ovington flew the first airmail letter for the U.S. Post Office Sept. 11, 1911. Louis Pacent became a leader in short-wave radio manufacturing and in the development of sound systems for motion pictures.

[10] *Electrical Review* (N.Y.), Nov. 9, 1898, p. 294.

[11] New York *Journal*, Nov. 13, 1898.

[12] *Electrical Review* (N.Y.), Nov. 16, 1898, p. 310.

[13] *The Sun* (N.Y.), Nov. 21, 1898.

[14] *Electrical Review* (N.Y.), Dec. 7, 1898, p. 365 (translated from *L'Événement*).

I shall be in Europe a year, yet.

Here in the hotel the other night when some interested men were discussing means to persuade the nations to join with the Czar & disarm, I advised them to seek something more sure than disarmament by perishable paper contract — invite the great inventors to contrive something against which fleets and armies would be helpless & thus make war thenceforth impossible. I did not suspect that you were already attending to that, & getting ready to introduce into the earth permanent peace & disarmament in a practical & mandatory way.

I know you are a very busy man, but will you steal time to drop me a line ?

Sincerely yours,

Mark Twain[15]

Need for secure control

It became apparent to Tesla that some means must be provided for secure control of remote devices in order to prevent interfering signals caused by natural effects, such as lightning sferics, or the deliberate intervention from harmful sources. Although the remote control of mechanisms was indeed an achievement, the matter of noninterfereable control now became essential to its practicality. He would later refer to this as the art of individualization, a technique for producing a control signal that resulted from the conjoint action of two (or more) wireless AC signals. This was a most ingenious idea.

Tesla's "individualization" concept

In a later article,[16] Tesla described his concept as follows:

[15] From Library of Congress microfilms of correspondence in archives of the Tesla Museum, Belgrade, acquired in 1961, consisting of correspondences between Tesla and Samuel Clemens (Mark Twain), Robert Underwood Johnson, J.P. Morgan (Sr. & Jr.), George Scherff, George Westinghouse, and the Westinghouse Electric & Manufacturing Company.

[16] Tesla, in Part VI - "The Art of Telautomatics" (Introduction, n. 2.).

> The underlying principle can be explained in a few words. Suppose that there are n simple vibrations suitable for use in wireless transmission, the probability that any one tune will be struck by an extraneous disturbance is $1/n$. There will then remain $n-1$ vibrations and the chance that one of these will be excited is $1/n-1$, hence the probability that two tunes would be struck at the same time is $1/n(n-1)$. Similarly, for a combination of three the chance will be $1/n(n-1)(n-2)$ and so on. It will readily be seen that in this manner any desired degree of safety against the statics or other kind of disturbance can be attained provided the receiving apparatus is so designed that its operation is possible only through the joint action of all of the elements. . . .

Using this analysis with n equal to 100—a practical value at the time this invention[17] was conceived—and a combination of three frequencies (vibrations) selected, the chance that these three frequencies would randomly occur is one in a million.

Much of the experimental work at Tesla's laboratory on Houston Street in New York City involved high-voltage resonators capable of producing discharges measuring 16 feet end to end—the limit with discharges touching the walls. Therefore, he temporarily left that laboratory in mid-1899 to establish an experimental station in Colorado Springs where he would not be confined by the limits imposed in New York. En route to Colorado Springs, Tesla gave a lecture before the Commercial Club at the Auditorium in Chicago May 13th and again demonstrated the wireless control of his boat, previously shown at Madison Square Garden, from the shore of a small lake.[18, 19]

Tesla had visited Colorado Springs before—once in 1896 to conduct curious experiments in radio transmission and reception along the rise of Pikes Peak,[20] and again in January of 1899 to

[17] U.S. Patent No. 613,809, Nov. 8, 1898, lines 271-277. (See Appendix A.)

[18] "Tesla's Lecture in Chicago," *Electrical Review* (N.Y.), May 17, 1899, p. 309.

[19] "Tesla's visit to Chicago," *Western Electrician*, May 20, 1899, p. 285.

[20] "Earth Electricity to Kill Monopoly," *The World* (N.Y.) Sun. Magazine, Mar. 8, 1896, p. 17.

choose the site for the experimental station.[21] Among the very first experiments conducted by him following construction of the experimental station, only six weeks after his arrival in Colorado Springs, was conclusive verification of his individualization invention by wireless, which had only been tested by induction at his laboratory in New York City.

Tesla's patent application for the art of individualization was filed after his return to New York July 18, 1900, Serial No. 23,847, with the title, "Systems of Signaling." The application was subsequently divided by Tesla and an application for "Method of Signaling" filed June 14, 1901, with this application of the pair assigned Serial No. 64,522. Nearly a year passed. Then, a possible conflict in Tesla's "Systems of Signaling" application with an application filed by Reginald Fessenden was determined by the U.S. Patent Office. An Interference notice dated April 22, 1902, was forwarded to both applicants. The determination on the Interference would, of course, affect both of Tesla's applications. The case was resolved in Tesla's favor January 26, 1903, with the "Method of Signaling" patent granted March 17, 1903, as No. 723,188, and the "System of Signaling" patent ("Systems" changed to singular) granted April 14, 1903, as No. 725,605.

It should be noted that the Wilson and Evans patent, "Method of Controlling Mechanism by Means of Electric or Electromagnetic Waves of High Frequency,"[22] shows a glimmering of an idea for security control by coincidental transmission of a signal by means of two mutually perpendicular short-wave Hertzian dipoles, thereby obtaining a pseudo two-channel transmission for conjoint detection. However, in such an arrangement, the single transmission "channel" would be subject to interferences, and the patent carries no claim of a method for carrying out the invention or any allusion to security advantages for the method.

Recognizing the U.S. Navy's critical interest in defense against torpedoes from the demonstrations at Madison Square Garden of

[21] From Library of Congress microfilms of correspondence in archives of the Tesla Museum, Belgrade (see n. 15); letter dated Jan. 4, 1899, from Scherff in New York City to Tesla at the Alta Vista Hotel, Colorado Springs.

[22] See n. 3.

other experimenters' ideas shortly before his own private demonstration there of a radio-controlled boat, on October 19, 1900, Tesla communicated with the Navy Department as to whether tenders were being made for wireless telegraphic apparatus, as well as for the operation and control without wires of submarine torpedoes or other vessels, and if so, he requested an opportunity to demonstrate his inventions.[23]

Lieutenant (afterward Captain) Richmond Pearson Hobson, who became a hero of the Spanish-American War with the scuttling of the *Merrimac* in the harbor of Santiago, was introduced to Tesla at a dinner by the wife of Robert Underwood Johnson, editor, *Century* magazine. Mrs. Johnson wrote to Tesla,

> Will be delighted to dine with you Saturday evening next. Will you please give us the hour? We are having with us at present Lieutenant Hobson, a very nice young fellow he is, by the way, and of course you will want us to bring him with us.

They became close friends and Hobson, interested in Tesla's inventions, on May 6, 1902 wrote,

> I have brought the matter of your inventions in wireless transmission of electrical energy in connection with the naval exhibit at Buffalo to the attention of the Navy Department and you will get an official letter on the subject. . . . Do send something, if only plans, at once, that is before the opening of the 29th inst. . . . and follow same by the craft.
>
> Should you find it impossible to complete it in time, why could you not send the model I saw in you laboratory in '98 and your patent drawings and descriptions, even if you could send nothing else?
>
> Do not hesitate to write freely to the Chief Clerk, Mr. Peters . . . who is very cordial about the idea of including your patents in our exhibits. . . . I think this is a good opportunity for bringing you patents to the attention of the Navy without the usual difficulties of formalities. I think these patents have great

[23] U.S. Navy Department records.

value for our Navy and country and therefore my dear Tesla, do not fail in this matter of the first step toward their introduction.

R.P. Hobson

A later communication, undated, refers to the above. Hobson tells Tesla there has been a fight within the Navy about Tesla's wireless exhibit being shown in the Navy Department Exhibit, the result of a feud not directly related to the invention but rather to something ongoing between two high officers. This coincidentally resulted in the rejection of Tesla's craft (the remote-controlled boat).[24]

This was not the first approach made by Tesla to the Navy Department for the possible utilization of his inventions and was far from the last.

In the meantime, on October 23, 1900, a pair of means-and-method patents[25] was granted to Lieutenant (later Admiral) Bradley A. Fiske broadly covering the wireless control of mechanisms. However, Fiske's application was dated September 7, 1898, two months after Tesla's application and two months before the patent was granted to Tesla. The U.S. Patent Office failed to take notice that two applications were at hand for essentially the same invention. An Interference should have been declared before the

[24] Copies of these letters were obtained by Kenneth Swezey, close friend of Tesla for 20 years to the time of death, during his 1957 visit to the Tesla Museum in Belgrade.

[25] Nos. 660,155 and 660,156, applications filed Sept. 7, 1898. "Single patents having both method and means claims are quite common today as well as in the past. On the other hand, separate patents for method and means inventions seem to be . . . more common in the past than in the last several decades. Both methods and means for carrying out the method have been statutory subject matter of invention since the beginning. Multiple patents have become less popular. Prosecuting two patent applications is obviously more expensive than prosecuting only one. On the other hand, there may have been certain advantages for the patentee in having separate method and means patents. Or, it may be that the Patent Office itself promoted separate patents for administrative or fee-gathering reasons! It's an interesting question, and the fact that there is no ready answer to it illustrates how difficult it will ever be to understanding the history of patenting in the U.S. without a comprehensive, functional history of the Patent Office." From a private communication from Kendall J. Dood, U.S. Patent Office examiner, retired, dated Dec. 26, 1995.

patent was granted to Tesla. By an arrangement with the Western Electric Company (assignee of the application), the Fiske patent was granted. Tesla was not able to prove invention before Fiske, and Fiske therefore obtained the primary claims on the concept of wireless control of mechanisms although the Tesla patent was in place. The reasons Fiske's name is not more prominently mentioned in the history of radio-controlled devices are: (a) he did not construct and exhibit an operative craft as Tesla, and Wilson and Evans had done; (b) in terms of patent application date, Wilson and Evans preceded him; and (c) the issue of security control of such craft, represented by Tesla's art-of-individualization invention, now known as the fundamental AND logic gate (described later), took precedence over the invention concept of wireless remote control.

What did survive in the Tesla patent are the methods of signal detection. A number of writers[26] have taken note of the ingenious decoherence method disclosed by Tesla in Claims 9 and 10 of his radio-control patent (No. 613,809).

Later contenders

The U.S. Government had begun to turn serious attention to reliable radio-guidance technology. John Hays Hammond, Jr., scion of perhaps the most famous mining development engineer of the time, became interested in the work of Bell, Edison, and Tesla, especially on the idea of utilizing the latter's ideas for remote control of devices by wireless (radio). While a student at Yale and editor of *Yale Scientific Monthly*, he was probably surprised that Tesla had recently contributed an article to *Harvard Illustrated Magazine*.[27] Not to be outdone by Harvard, he wrote to Tesla asking if he would have time to prepare an article on remote control by radio for the university magazine. Commencement was near at hand (1910), and Hammond was offered a position of editor

[26] Among them, at widely separated times, Sewall, Charles H., *Wireless Telegraphy; Its Origins, Development, Inventions, and Apparatus*, D. Van Nostrand, New York, 1904, p. 153; and Lynde P. Wheeler, "Tesla's Contribution to High Frequency," Pt. II of "Nikola Tesla's Achievements in the Electrical Art," *Electrical Engineering*, Aug. 1943 (v. 62), pp. 351-357, ref. at p. 356.

[27] "Signaling to Mars—A Problem in Electrical Engineering," Mar. 1907 issue.

for *The Electrical Age*. Tesla replied October 13, 1910, after Hammond had left Yale, that he was not able to comply earlier and inquiring if he was yet interested in such an article. Because of Hammond's departure from Yale, the article never materialized.

Supported in the idea of radio guidance by his father, a laboratory was established for Hammond in a bungalow on the estate grounds some distance from family castle—afforded with a drawbridge and moat—on Lookout Hill off Gloucester Bay. His father would keep the money flowing as long as there was progress.

By that time, Tesla, always in search of working capital for his research endeavors, had made acquaintanceship with Hammond Sr. —said to have influenced the nomination and election of President Taft and was also the President's special representative at the coronation of King George V—concerning investment funding for his spiral-flow disk turbine. Hammond Jr. began making public statements on his plans in radio guidance development and Tesla took particular notice. The following amusing exchanges ensued:

November 8th, 1910

My Dear Mr. Hammond:

I was glad to read the enclosed newspaper reports. This is water on my mill. Just go ahead and make a lot of money, then I will sue for infringement and we will divide.

Yours sincerely,

N. Tesla

* * * * *

November 10th, 1910

My dear Mr. Tesla:

It will be very agreeable for me to share profits with you, but I shall only do so on the condition that you share our liabilities also...

The Yale faculty and undergraduates are endeavoring to get me to use what small influence I may have to induce you to deliver a lecture at Yale some time during the month of February. Having just graduated myself, I know the feeling that is prevalent in New Haven and can assure you that they would appreciate greatly having the pleasure of hearing you.

Yours sincerely,

John Hays Hammond, Jr.

* * * * *

November 12th, 1910

My Dear Mr. Hammond:

Replying to your letter of the 10th, I am very glad to know of at least one modest inventor. As I naturally surmise that your Papa would pay all our liabilities, I am willing to share in these.

I shall look to the pleasure of seeing you on your next visit to New York, when you will have an opportunity of seeing a magnificent little pump in my office.

Yours sincerely,

N. Tesla

* * * * *

November 14th, 1910

My Dear Mr. Hammond:

Judging from the enclosed, I think that you are playing a wireless possum. Notwithstanding your assurances, I will watch your progress and bring a friendly suit for infringement as soon as I ascertain that you are in funds.

Yours sincerely,

N. Tesla

A year passed. Hammond had dinner with Tesla in New York and the following letter was received from Hammond after his return to Washington:

1500 Rhode Island Avenue,
Washington, D.C.
February 16th, 1911

Mr. Nikola Tesla,
Metropolitan Tower, New York.

My dear Mr. Tesla:

I have just had a most interesting conference with the leading government authorities in the Wireless Department, and they seem to think that the field I am working in is a most valuable one for naval work.

As you know, I am endeavoring to develop a short-distance non-interferable system for use in fleet action. I will have a 5 K.W. sending station and a flat-top aerial 118 feet high. Only distances up to 20 miles need be covered, and transmission of 20 words per minute is the maximum requirement, written message on Morse tape preferable. These are the Government requirements. Of course the great difficulties we have to contend with are forced oscillations, and atmospheric disturbances, neither of which can be eliminated by known methods of tuning, as neither Marconi's nor Fessenden's "Interference Preventive" seems to have been successful. By using a lot of power and a sensitive detecting device, with a stiff receiving circuit having a very high inductance or very loose coupling, I think that we can eliminate much outside interference.

On the strength of our conversation at dinner the other night, I have become much interested in your idea of our forming a joint company such as the "Tesla-Hammond Wireless Development Company." (In thinking of this name I have followed Emersonian advice, and, as you see, attached my chariot to a Star.) The purpose of this Company would be to perfect an automatic selective system, to perfect the dirigible torpedo, and eventually to carry out your magnificent projects that will wirelessly electrify the world. My own endeavor up to the present time in the development of some practical dirigible torpedo apparatus will not be valueless, and my imagination has made me a believer and a devotee in the art of which you are the High Priest -- "Telautomatics." It is on these considerations that I think a little, unpretentious company should be energized which may further the seeds of great possibilities.

On my return to New York I will call and see you and show you my idea for a selective system. If this idea of a small company to protect and develop our wireless patents appeals to you in the broad outline I have given you, do drop me a line here, and after having your consent I will communicate with my brother Harris and describe to him our plan.

Hoping that I am not monopolizing too much of your valuable time,

I am most sincerely yours

John Hays Hammond, Jr.

Tesla was apprehensive, and replied as follows:

February 18th, 1911

My Dear Mr. Hammond:

I was glad to receive your letter of the 16th and to know that the Government officials are waking up to the importance of your efforts. The exposition of your selective system is looked to with great interest. It would just break my heart if it should turn out that my own selective system is better than yours, notwithstanding the fact that I know it would make you happy for I have discerned that you are a gallant fellow.

The Tesla-Hammond combination looks good to me but we should have to go at it with some circumspection. I have already interested a gentleman who signs himself J.P.M. in a part of my wireless inventions and my friend Astor is now waiting for the completion of my plant to go into the wireless power transmission business which should be a colossal success. In the art of Telautomatics, however, I am perfectly free and would be glad to go into any fair proposition to exploit the field. I think that in a few years this departure will command the attention of the world.

I have just completed my turbines and am starting Monday to install them at the Edison plant where I expect to show them to you in operation on your next visit to the city.

With kind regards,

Very sincerely yours,

N. Tesla[28]

The government's interest languished. Then, in an article appearing in the *National Press Reporter* titled "The Future in Wireless,"[29] Hammond acknowledges the discovery and inventions of Tesla in radio communication but goes on to say it "was left to the splendid enterprise and indefatigable endeavor of Marconi to crystallize the results of previous investigations. . . ."

This brought a prompt response from Tesla pointing out, among other things, that Marconi's apparatus is in every important particular his own.[30] The blush now seemed off any Tesla-Hammond joint enterprise.

Fritz (Frederick) Lowenstein, who had worked for a time at Tesla's Colorado Springs experimental station and gave one of the depositions in the U.S. Patent Office Interference case on Tesla's art-of-individualization invention, was a consulting engineer to the U.S. Navy and Hammond's projects in 1911. He introduced Hammond to Benjamin Franklin Miessner who had completed a "hitch" in the Navy and already obtained a significant radio detector patent. When in the Navy, Miessner met many of the early radio pioneers including Lowenstein, Louis W. Austin, George H. Clark, Reginald Fessenden, Walter C. Massie, Greenleaf Whittier Pickard, John Stone Stone, and David W. Todd. Ham-

[28] From correspondences placed in the Library of Congress by executors of the estate of J.H. Hammond, Jr.; photostats of some correspondences were also made and placed in the collection of the Engineering Societies Library, New York City.

[29] May, 1912 (Vol. 15, No. 10).

[30] Tesla letter to Hammond, May 22, 1912.

mond was favorably impressed with Miessner and extended an invitation to work at the Gloucester laboratory.

Miessner was first assigned to examine the Marconi approach to torpedo guidance equipment with his engineers in New York and thought the concept "preposterous." Nevertheless, he was instructed to have it shipped to the Gloucester laboratory. Tests were made on the bay but they failed miserably. Hammond became despondent and scrapped the Marconi apparatus. Hammond Sr. was fully aware of the failure and sent for Miessner to have a private discussion for advice whether to continue to fund his son's activity. Miessner advised that the project was certain to succeed by adopting *his* approach to the problem and saying if Hammond Sr. would give him another year, he would show him a successful surface-torpedo control system. Much to Miessner's surprise, Hammond Sr. took him up on it.[31]

An important visitor came to the Gloucester laboratory in the fall of 1912—Lee de Forest. As Miessner describes the visit,

> [W]e had a mutually edifying session, each party trying to find out what the other had been up to. I was quite free with my disclosures, and explained to him my vacuum-tube amplifier and detector systems and described my selective systems and apparatus. I also asked him some leading questions.[32]

Miessner also worked at the Lowenstein laboratory as his half-time assistant. Lowenstein cautioned him to keep careful records of the work he was doing (to avoid any conflict of interest problems that could arise). In the meantime, there began to develop a competitiveness between Miessner and Hammond over credit and title to certain inventions. A stormy session ensued. Miessner left the Gloucester laboratory in November (1912) and returned to his Indiana home entering Purdue University.

It would appear that Hammond recognized Tesla's art-of-individualization invention was now crucial to the success of his efforts

[31] Miessner, B.F., *On the Early History of Radio Guidance* (San Francisco Press, 1964), p. 6.

[32] Miessner, B.F., p. 32,

in secure guidance control. He at first attempted to dodge the Tesla patent by other cumbersome modifications, be that as they may patentable, but these failed. Development then proceeded at the Gloucester laboratory in 1914 using a form of the Tesla dual-channel type control for the experimental houseboat *Natalia*, moored near Hammond's sailing yacht.[33] What is a matter of conjecture, but most probable considering the circumstances and individuals, Hammond Sr. exercised his formidable political influence to obtain a 1916 Government appropriation in the amount of $750,000 for Hammond Jr. Of it, Tesla remarked,

> I tried to persuade the Navy. I put it before capitalists with a view of introducing it, but it was absolutely impossible to find listeners until my patent expired, and now Congress has appropriated $750,000! My patent has expired—I get nothing, and a much smarter man, but of much lesser experience than my own, is to get $750,000. I wish him luck. But, still, I ought to have had something, I believe.[34]

Lee de Forest was not so reserved.

De Forest to a Congressman, June 13, 1916:

> On April 20th I wrote you at length to point out just why the proposed gift to John Hays Hammond, Jr., of $750,000 in Government funds for "rights to certain pending patent applications" was a gross extravagance and 'paternalism' of the worst sort. . . .
>
> When this nation is willing to pay $750,000 for an arm of defence [*sic*] whose chief merit is to act as a bunk scarecrow we will deserve the fate that then awaits us. . . .[35]

While an undergraduate, the manuscript for Miessner's book,

[33] Hammond, J.H., Jr., and E.S. Purington, "A History of Some Foundations of Modern Radio-Electronic Technology" (see n. 2).

[34] Anderson, L., ed., *Nikola Tesla On His Work With Alternating Currents . . .*, p. 158.

[35] Miessner, B.F., pp. 39, 40.

Radiodynamics,[36] was completed. He began the manuscript during his employment at the Gloucester laboratory. The publisher's reader thought the manuscript well done but felt that some statements may sound "over sanguine" to some readers. Miessner remarked,

> I am almost certain that the reader who reviewed the book for Van Nostrand was none other than Nikola Tesla. Since I was planning to include descriptions and drawings of his pioneering efforts in the field of automatic control, I had written to him, only to find that he was already well acquainted with my project. We exchanged three letters in all.[37]

The first reply from Tesla, September 29, 1915, is almost entirely quoted in Miessner's *On the Early History of Radio Guidance* (pp. 43, 44). It is somewhat amusing to find that, by convenient ellipsis, the last sentence in the first paragraph criticizing the title, *Radiodynamics*, is omitted.

Hammond demonstrated accurate control of the surface boat *Natalia* to the Navy Board August 23, 1918. The Board adjourned without advising Hammond of its findings. Two months passed and Hammond was notified that the Board could not make a favorable ruling, indicating a requirement for the control of a completely submerged torpedo.

Hammond protested the revised requirement but, on the belief that the War Department would ultimately cover his expenses, continued work. On July 29, 1921, the officer in charge of the project reported that the tactical value of the weapon had been almost entirely lost in view of recent aircraft bombing tests and recommended discontinuance of the Hammond project.[38]

[36] See n. 1.

[37] See Appendix G. This editor was shown Tesla's replies at the flat of Kenneth Swezey in the Greenpoint neighborhood of Brooklyn in the 1950s. Miessner had given them to Swezey who was developing a collection of notable pioneers inspired by Tesla's work. The letters were acquired by the N.Y. Public Library, with photostat copy sets deposited in the Library of Congress, in 1973 following Swezey's death.

[38] Howeth, L.S., *History of Communications-Electronics in the United States Navy*, U.S. Government Printing Office, 1963; pp. 337-342.

The guidance of a torpedo by radio under water is technically much more difficult than the guidance of a craft, or weapon, at the surface or in air. This arises because of the large conductivity of seawater. The attenuation of signals propagated from a coast station through seawater to a thin cable trailing a submarine would be enormous. A 1960s project of the U.S. Navy in Wisconsin called "Sanguine" (later reduced in scale to project "Seafarer" in Michigan) involved a grid of several thousand miles of buried antenna to propagate ELF (extra-low frequency—30 to 300 Hz) signals in the earth-ionosphere waveguide. Although signal penetration in seawater is attenuated exponentially by depth and frequency, with such low frequencies detection at several hundred feet is possible.[39] The Hammond test was conducted at 188 kHz.

The historic paper[40] by Hammond and Purington 35 years later on their role in the development of radio guidance erupted in a major critique by Lloyd Espenschied—historian and co-inventor of coaxial cable—further delineating the work of Lowenstein, Miessner and Tesla, among other issues, through rebuttal and surrebuttals.[41] Having been 'stung' by his experience with Hammond, Miessner became an advocate of independent inventorship and outspoken on the plight of captive corporate inventors. He submitted a profusion of communications on the rare gift of individual creativity, criticizing the notion of team "brain storming" and "piracy" of an employee's inventive output as a condition of employment on government contracts.[42] His book, *On the Early*

[39] Wait, J.R., "Propagation of ELF Electromagnetic Waves and Project Sanguine/Seafarer," *IEEE Journal of Oceanic Engineering*, Apr. 1977 (Vol. 2, No. 2), pp. 161-172. Wait refers to investigations of "Tesla the Prophet" having a striking similarity to the Navy projects.

[40] Hammond, J.H., Jr., and E.S. Purington, "A History of Some Foundations of Modern Radio-Electronic Technology" (see n. 2), p. 1185.

[41] Espenschied, Hammond, and Purington, "Discussion of 'A History of Some Foundations of Modern Radio-Electronic Technology,'" *Proceedings of the IRE*, July 1959, pp. 1253-1268; editorial comment by J.D. Ryder, p. 1189.

[42] Among them, "On the Statistics of Individual Variations in Productivity in Research Laboratories," *Proceedings of the IRE*, Oct. 1957, pp. 1409, 1410; "Credit to wartime inventors," *Radio-Electronics*, July 1961, p. 45; "Creativity and Inventions." *Electronics*, Sept. 14, 1962, p. 4; ". . . companies and inventors," *International Science and Technology*, April 1963, p. 13; "Government Patent Policy" and "Patent Rights Law," *Electronics*, May 3 and 10, 1963, p. 4.

History of Radio Guidance, was an attempt to finally "set the record straight" on the Hammond-Purington article. Highly relevant information is presented on the title topic including behind-the-scenes events surrounding inventions of the audion, superheterodyne systems, and frequency modulation.

Guided weapons

The technical obstacles associated with, and diminished tactical need for, *radio* control of torpedoes eliminated them for all practical purposes from the military family of arsenal weapons. However, the possibilities of radio control of other devices—surface vehicles, aircraft, and missiles—would later capture military attention.

Tesla's telautomaton was conceived as an engagement craft in warfare. Distressed by the losses in war, he and others envisioned that some weapon could be developed to make war unthinkable. World War I had gripped the nation and concern over it was evident in the daily press years following. In an interview article, Tesla provided captioning for a scene of war of the future:

> War of the future as it will be conduced from the viewpoint of Dr. Tesla. – Machines of destruction more terrible than anything concocted by the master minds behind the "World War." Armies and navies will sail under the ocean and through the skies with not a man on board. . . . These death-dealing monsters of the sea and air will be controlled and directed from distant points hundreds or even thousands of miles away by radio waves of the proper sequence and frequency. . . . Man will be the master mind behind the future war, but machines only will meet in mortal combat. It will be a veritable war of "Science."[43]

We find expressed here, five decades earlier, many of the same technologies now deployable.

[43] Caption for drawing by artist Frank Paul accompanying "An Interview With Nikola Tesla" by H. Winfield Secor, *Science and Invention*, Dec. 1921, pp. 912, 913, 957-959; illus. on p. 912.

Man's most far-reaching scientific discoveries and technological achievements have been measured at times of threat to a nation's security. When Hiero, tyrant of Syracuse, ordered a crown of pure gold, he asked consultant Archimedes to determine whether any other metal had been used. Archimedes' solution to the problem is the classic tale of him running down the street shouting *Eureka, Eureka!* But, far removed from satisfying mundane orders, Archimedes also devised war machines that Syracuseans used against the Roman legions. Leonardo da Vinci also served under Pope Alexander VI as engineer general of the army during the war in Central Italy. As Robert Duncan, Chief, Guidance and Control Divison, U.S. Air Force observed,[44]

> The search for knowledge and understanding has always been under stress. . . . The ancient mathematician manipulating his triangles in a quest for a release from his perplexity by logic was was never far from the barbarian and his spear. A shadow of the guillotine fell across the page on which France recorded some of its most profound science. No man delves into the unknown who is not under sentence of death.

Although Tesla conceived his telautomaton invention as a weapon of war, he also recognized the broad application of remote-controlled devices in other environments. The last paragraph of the patents's description reads:

> The invention which I have described will prove useful in many ways. Vessels or vehicles of any suitable kind may be used, as life, dispatch, or pilot boats or the like, or for carrying letters, packages, provisions, instruments, objects, or materials of any description, for establishing communication with inaccessible regions and exploring the conditions existing in the same, for killing or capturing whales or other animals of the sea, and for many other scientific, engineering, or commercial purposes. . . .

Today we find all of these applications implemented—small vehicles directed into hostile environments such as poisonous atmospheres or standoff police engagements, and reconnaissance

[44] Keynote address, Third Biennial Inertial Guidance Test Symposium, Oct. 19, 1966, Holloman AFB, N. Mex.

aircraft. It would seem that space agencies throughout the world have squandered enormous revenues of their respective countries pandering to man's ego by thrusting him into the great hostile environment of outer space. Resources expended by these nations might better serve the development of non-manned space vehicles adequately controlled from earth. An increasing number of scientists and analysts, including Edward Hodgins and the late Carl Sagan, have called for the cancellation of the space station, phasing out of shuttle flights, and overhaul or termination of the human space program utilizing instead satellites and robotic technology.[45] The issue of secure remote control remains the same but more difficult given the plethora of emissions now encountered. Refined approaches having their roots in Tesla's spread spectrum techniques, frequency-hopping and orthogonal frequency division multiplexing, as well as more recently developed deinterleaving, data fusion, and super-resolution procedures are presently allowing desired data signals to be isolated from a jumble of intentional and unintentional emissions. Most of these techniques, either directly or indirectly, depend upon comparisons or conjoint conditions of some type to produce the desired action. With respect to 'frequency hopping,' it is interesting to note that actress Hedy Lamarr played a significant roll in the conception of a switching device to make Tesla's original scheme possible. Together with avant-garde American composer George Antheil she patented a technique, using player piano rolls for programming of the switching sequence.[46]

Thus we find that Tesla was the father of both the guided weapon and secure remote-control technology on which the former is critically dependent in effective deployment. This was affirmed in a postscript to a letter this editor received from Edwin H. Armstrong 12 weeks before his death, reproduced below:

[45] Merzer, Martin, and Phil Long, "NASA: Glamour Is Gone, and Future is Bleak—Some scientists want satellites and robots," syndicated special report of Knight-Ridder News Service, June 8, 1995.

[46] U.S. Patent No. 2, 292,387, "Secret Communication System," application field June 10, 1941, granted Aug. 11, 1942, to Hedy Kiesler Markey (Lamarr) and George Antheil. See: Kahn, David, "Cryptology and the origins of spread spectrum," *IEEE Spectrum*, Sept. 1984, pp. 70-80, and Braun, Hans-Joachim, "Advanced Weaponry of the Stars," *Invention & Technology*, Spring 1997, pp. 10-16.

Columbia University
in the City of New York
[NEW YORK 27, N. Y.]
DEPARTMENT OF ELECTRICAL ENGINEERING

November 16, 1953

Mr. Leland I. Anderson
127 Seymour Avenue Southeast
Minneapolis 14, Minnesota

Dear Mr. Anderson:

Very truly yours,

Edwin H. Armstrong

Edwin H. Armstrong

EHA:ecw

P.S. In my "Appreciation" of the work of Tesla which came during wartime, there was much research under way toward reducing the Tesla concept of the guided weapon in various laboratories - I considered it inadvisable to draw any attention to the subject by reference to Tesla's early concept of the matter. So far as I know, the credit for the concept rests entirely with him.

EHA

E. H. A.

The AND logic gate

In 1891, Tesla patented the resonant coil system named after him.[47] Charles Steinmetz and Elihu Thomson, in their contemporary discussions of its operation, had not recognized the enhanced voltage buildup property of the coil's construction resulting from standing waves established in the secondary. Forty-four years later the inadequacy of existing mathematical treatment to describe the coil's operation was realized.[48] The magnifier principles that Tesla developed to produce extremely high voltages at kiloampere levels with this coil system are only now becoming understood.[49] It is interesting to note that the first photograph of Tesla's magnifier in action was published, without explanation as such, in 1898—see page 12 of the Interference transcript.

From his work in Colorado Springs, Tesla offered photographs of electrical effects, published soon thereafter, that have left those pursuing his writings astonished at the results obtained. Over 90 years ago, Tesla produced the greatest length electrical discharges achieved by man. Following his death in 1943, a record of these Colorado Springs notes was found and published shortly after his 120th anniversary.[50]

At this point, the reader may very well ask, what has all this to do with the invention of the AND logic gate? The question naturally arises because today one thinks in terms of digital computer semiconductor arrays operating at 3- to 5-volt levels. Modest Tesla coils are capable of producing voltages from a quarter of a million volts and up. But, in the early evolution of the idea, the ANDing function was demonstrated using Tesla coils.

[47] U.S. Patent No. 462,418 of Nov. 3, 1891, "Method and Apparatus for Electric Conversion and Distribution," application filed Feb. 4, 1891.

[48] Sloan, David H., "A Radiofrequency High-Voltage Generator," *Physical Review*, Jan. 1, 1935 (Vol. 47).

[49] Corum, James F., and Kenneth L. Corum, "The Application of Transmission Line Resonators to High Voltage Power Processing: History, Analysis, and Experiment," *Proceedings of the 19th Southeastern Symposium on System Theory*, Clemson University, Clemson, S.C., Mar. 1987, IEEE Computer Society, pp. 45-49 (catalog # TH0180-0); and, Hull, Richard L., "Tesla Magnifier Basics," *Electric Spacecraft Journal*, Issue 10, pub. 1994, pp. 7-12.

[50] *Nikola Tesla—Colorado Springs Notes* (Introduction, p. xv).

A statement of an AND logic gate's operation is:

When *all* inputs to the gate are on, an output will be provided.

This function is represented in the following Logic Truth Table for a three-input gate.

INPUTS			OUTPUT
A	B	C	$A \wedge B \wedge C$
high	*high*	*high*	*high*
high	*high*	*low*	*low*
high	*low*	*high*	*low*
high	*low*	*low*	*low*
low	*high*	*high*	*low*
low	*high*	*low*	*low*
low	*low*	*high*	*low*
low	*low*	*low*	*low*

It is seen that of all eight possible combinations of three inputs, an output is provided *only* when they are all "high," or ON.

The other basic logic gate is the OR, and an output will be provided when *any* input to the gate is on. From these ANDs and ORs (and the complimentary gate functions of NANDs and NORs), all basic digital computer functions are performed. The implementation of the logic function may be accomplished in a number of ways depending upon the specific application.

Electronic Elements

The AND logic gate is schematically represented below for an electronic array employing passive diode circuit elements.

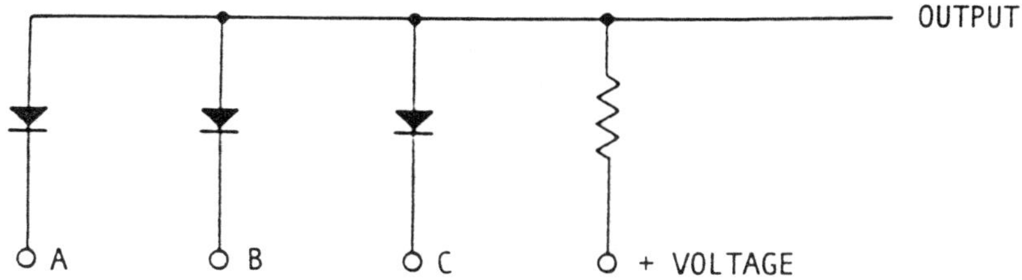

A + B + C INPUTS HIGH, OUTPUT HIGH

In this circuit, if *any* diode input is "low" the diode associated with that input will be conducting in the forward direction, indicated by the arrow symbol, pulling the output down to the low level. If, however, *all* the diode inputs are "high" (cut off), the output rises to the high level. The AND function is thereby accomplished when all the inputs are "high" producing an output "high." This is to be compared with the Logic Truth Table.

Reduced to their essentials, digital computer systems have thousands of logic AND, OR, and the complimentary NAND and NOR decision elements fabricated in very large semiconductor integration (VLSI) plug-in chip arrays. But these logic functions can be implemented by other than electronic means.

Non-electronic Elements

In the 1950s, concern was raised by the U.S. Department of Defense about the vulnerability of semiconductor computer elements to high radiation environments. It was recognized that ionization effects resulting from the detonation of a nuclear device could possibly render military computer systems inoperative. A great number of research and development efforts were undertaken to devise technologies 'hardened' against intense ionizing radiation.

One of the approaches used pulsed AC signals together with miniature ferromagnetic transformers. Logic functions were accomplished using a DC control winding to saturate the core of the transformer, thus selectively interrupting an AC signal path. Another approach spawned the development of fluid logic elements.

Fluid Logic Elements—The Tesla Turbine

In what initially may seem an unusual departure from semiconductor logic elements to fluid equivalents, we first turn to the disk turbine—a novel spiral-flow machine principle patented by Tesla nearly 75 years ago.[51] The turbine rotor consists of thin, closely spaced circular disks through which steam or other working fluid flows in a gradually decreasing circular path from the periphery to

[51] U.S. Patent Nos. 1,061,142, "Fluid Propulsion," and 1,061,206, "Turbine," of May 6, 1913, applications filed Oct. 21, 1909, and Jan. 17, 1911, respectively.

exhaust ports near the rotor's shaft. In extending the idea to expansive fluids, as a gas turbine, Tesla invented a 'valvular conduit' through which fluid flows easily in one direction but is greatly impeded in the reverse direction.[52] In a gas turbine application, fuel and air easily flow into the turbine's combustion chamber through two valvular conduits, but when ignition occurs, flow of the combustion gasses is greatly impeded in the reverse direction of the conduit and pressure is thus applied to the turbine rotor disks.

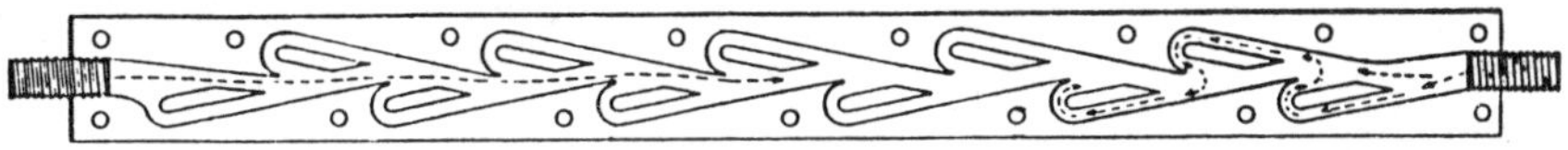

EASY FLOW DIRECTION ⇒

The Valvular Conduit

Tesla's valvular conduit has been described as a fluid diode[53]—the only mechanical valve invention having no moving parts. In functioning as a fluid diode, logic functions of ANDs and ORs can also be constructed from valvular conduit elements in micro-miniaturized form operating in a pulsed-fluid mode. Recently, engineers at the University of Washington created microchip-sized valvular conduits, etched on silicon wafers, for application in medicine, electronics, and other situations where small amounts of fluid can be pumped with no moving parts.[54]

High frequency, high voltage, conjoint oscillations demonstrating the AND function

When Tesla conducted experiments on dual-frequency transmission and reception at his Colorado Springs station using high voltage resonators, the confirmation of success was achieved by using instrumentation of his own design. Instruments to accurately

[52] U.S. Patent No. 1,329,559 of February 3, 1920, "Valvular Conduit," application filed Feb. 21, 1916.

[53] Wood, O. Lew, and Harold E. Fox, "Fluid Computers," *International Science and Technology*, Nov. 1963 (No. 23), pp. 44-52, 103.

[54] Paulson, T., "From an obscure genius comes a tiny pump for computer age," *Seattle Post-Intelligencer*, Oct. 16, 1996, pp. A1., A8.

measure frequency and phase were unavailable at that time. Today, experimenters and researchers are so accustomed to availing themselves of signal, spectrum, and time-domain reflectometry instrumentation that the thought of embarking on such experiments without them would be a severe obstacle.[55]

Imagine, then, Tesla facing this situation at his New York City laboratory. Realizing that he was entering a realm where measurement was the key to working with alternating currents, he set about to devise instruments to measure phase, frequency, and beats of two simultaneously generated frequencies. The description of these instruments was presented at his 1897 lecture before the New York Academy of Science.[56]

In reading accounts which followed the experiments he conducted at the Colorado Springs experimental station and the prototype facility at Wardenclyffe it is seen that Tesla, by means of these instruments of his own design, was able to adjust the frequency of oscillation of these enormous magnifying transmitters so that the quarter-wavelength potential maxima would be located precisely at the elevated terminal.[57] Remarkable for that time indeed.

Progenitor of the computer industry

The world's first high-speed electronic digital computer, ENIAC, was announced in 1946 by a University of Pennsylvania team under direction of John W. Mauchly and J. Presper Eckert.[58] Occupying a floor space of more than 15,000 square feet, it weighed more than 30 tons and used more than 18,000 electronic vacuum tubes. Ten years later, the first all solid-state computer was manufactured

[55] A 1987 Hewlett Parkard advertisement carries an illustration of Tesla amid sparks at his Colorado Springs experimental station with the caption, "Every day, the measurements get tougher . . ."

[56] Anderson, L., ed., *Nikola Tesla: Lecture Before the New York Academy of Sciences, April 6, 1897* (Breckenridge, Colo., Twenty First Century Books, 1994), pp. 58-69.

[57] U.S. Patent No. 1,119,732 of Dec. 1, 1914, "Apparatus for Transmitting Electrical Energy," application filed Jan. 18, 1902, renewed May 4, 1907.

[58] Their U.S. Patent No. 3,120,606 of Feb. 6, 1964, "Electronic Numerical Integrator and Calculator," application filed June 26, 1947, had a record 231 pages.

and delivered as a commercial product. The AND logic-gate technology employed by these machines was now recognized as Tesla's invention. Among the first to credit Tesla in this were Hammond and Purington.[59]

As well, in the mid-1950s, a graduate student at Iowa State University had traced high-speed electronic digital computer technology to the Atanasoff-Berry machine at the university, leading the Honeywell Company to challenge the ENIAC patent. A 1971 judgment found John V. Atanasoff as the true inventor. On November 13, 1990, at a White House ceremony, Atanasoff, then 87, was presented a National Medal of Technology.

A parallel redress occurred concerning the system patent for radio. Marconi was acclaimed early on, but Tesla continuingly asserted his priority and attempted to defend his patent claim. When the Marconi Company sued the United States for use of radio apparatus during WWI, the government took note of Tesla's assertion of priority and based its defense on the Tesla patent. The litigation that began following WWI was concluded, through several appeals, in 1943—six months after Tesla's death when the United States Supreme Court declared the Marconi patent invalid based on Tesla's earlier work and patent.[60]

Tesla could not have foreseen the development of the computer industry as we know it today. His invention of security for remote control of devices, or telemechanics, was not envisioned for what is recognized today as the fundamental AND logic gate in digital computer systems. But he was resigned to being far ahead of his time. In 1915 *The New York Times* carried a front page dispatch from Reuter's that Tesla and Edison were to share the Nobel Prize in physics. The prize was never awarded, and since that time biographers have speculated that Tesla refused the divided prize. Four days following the press story, in a handwritten note, Tesla strangely foretold the outcome:

[59] Hammond, J.H., Jr., and E.S. Purington, , "A History of Some Foundations of Modern Radio-Electronic Technology" (p. 125, n. 2), p. 1193: "Tesla had already proposed a security system based upon the coincidental transmission on two channels: a forerunner of the 'and' principle of modern computers."

[60] *United States Reports* (Vol. 320). "Marconi Wireless Telegraph Company of America v. United States," decided June 21, 1943, pp. 1-80.

In a thousand years, there will be many recipients of the Nobel prize. But I have not less than *four dozens* of my creations *identified with my name* in technical literature. These are honors real and permanent, which are bestowed, not by a few who are apt to err, but by the whole world which seldom makes a mistake, and for any of these I would give all of the Nobel prizes during the next thousand years.[61]

L.I.A.
Nov. 1, 1996

[61] From Library of Congress microfilms of correspondence in archives of the Tesla Museum, Belgrade (Part 2, n. 15); note to Robert Underwood Johnson dated Nov. 16, 1915.

Appendix

No. 613,809. Patented Nov. 8, 1898.

N. TESLA.

METHOD OF AND APPARATUS FOR CONTROLLING MECHANISM OF MOVING VESSELS OR VEHICLES.

(No Model.) 5 Sheets—Sheet I.

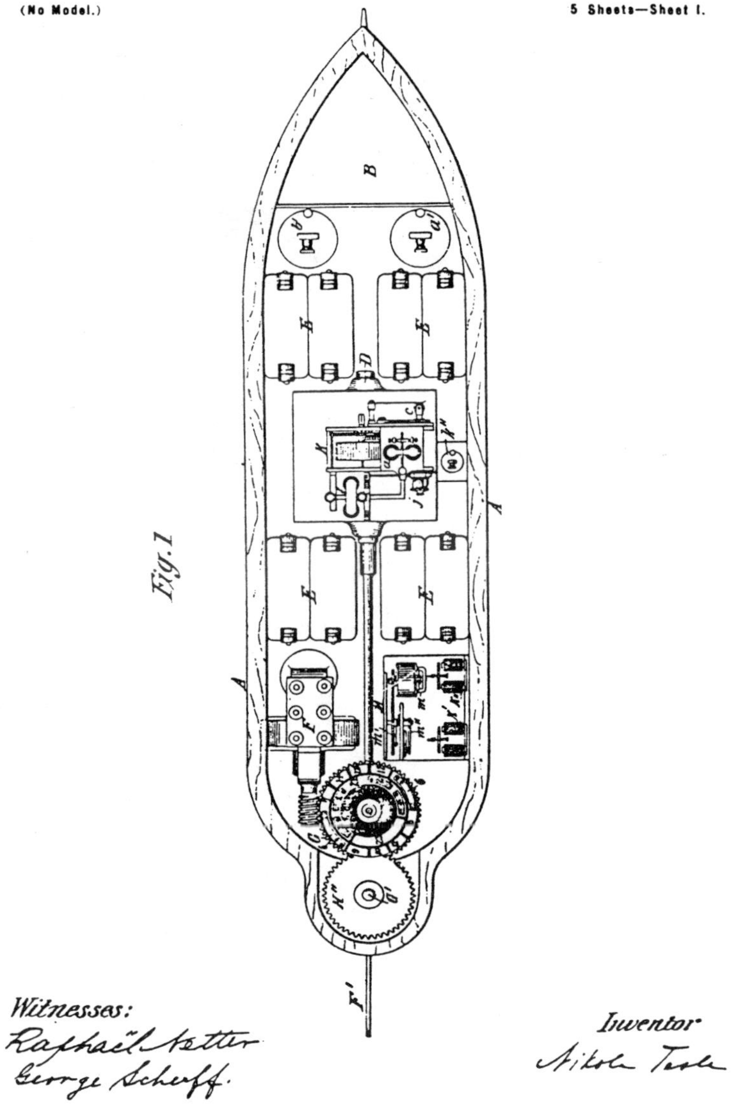

No. 613,809. Patented Nov. 8, 1898.

N. TESLA.

METHOD OF AND APPARATUS FOR CONTROLLING MECHANISM OF MOVING VESSELS OR VEHICLES.

(No Model.) 5 Sheets—Sheet 2.

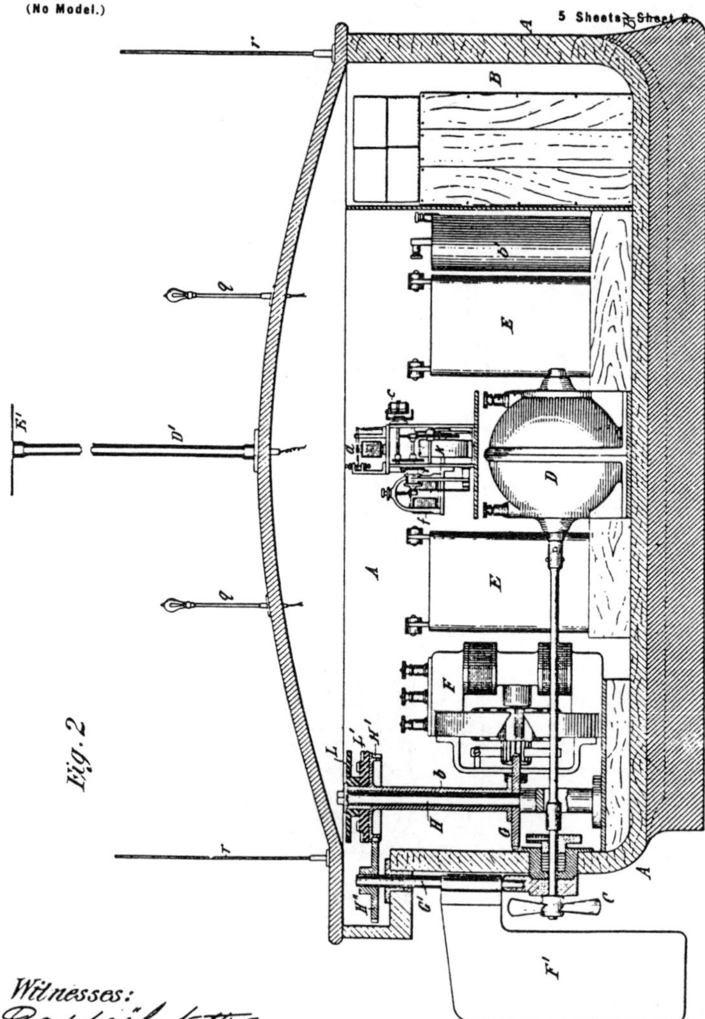

Witnesses:
Raphaël Netter
George Scherff.

Inventor:
Nikola Tesla

No. 613,809. Patented Nov. 8, 1898.

N. TESLA.

METHOD OF AND APPARATUS FOR CONTROLLING MECHANISM OF MOVING VESSELS OR VEHICLES.

(No Model.) 5 Sheets—Sheet 3.

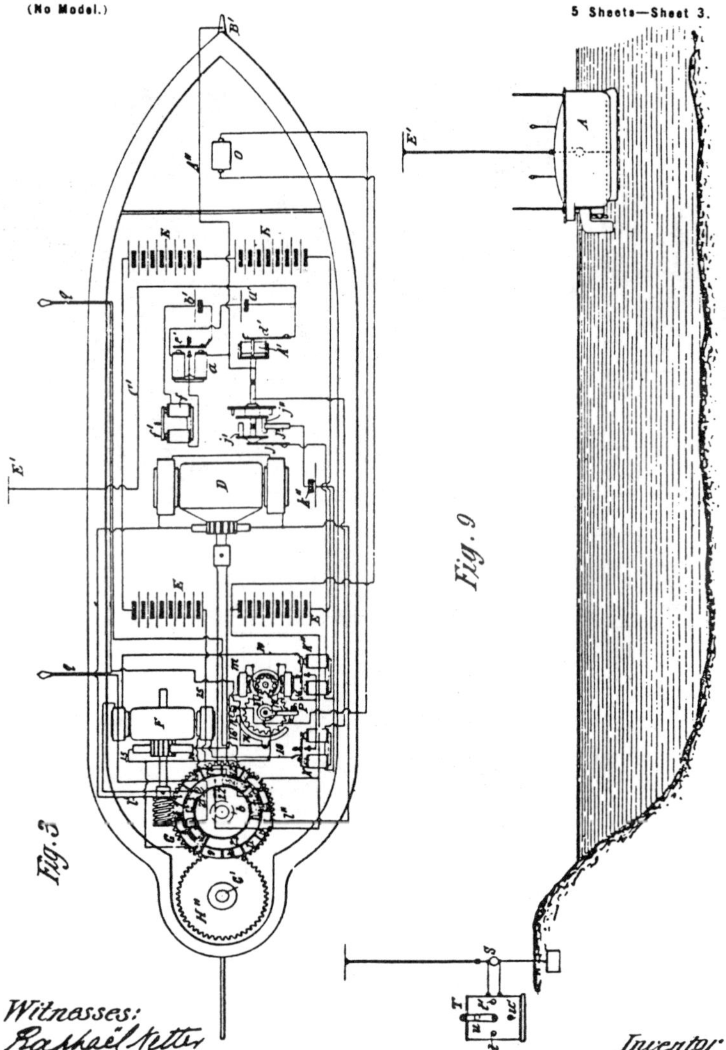

Fig. 3

Fig. 9

Witnesses:
Raphaël Netter
George Scherff

Inventor
Nikola Tesla

No. 613,809. Patented Nov. 8, 1898.

N. TESLA.

METHOD OF AND APPARATUS FOR CONTROLLING MECHANISM OF MOVING VESSELS OR VEHICLES.

(No Model.) 5 Sheets—Sheet 4.

Fig. 5

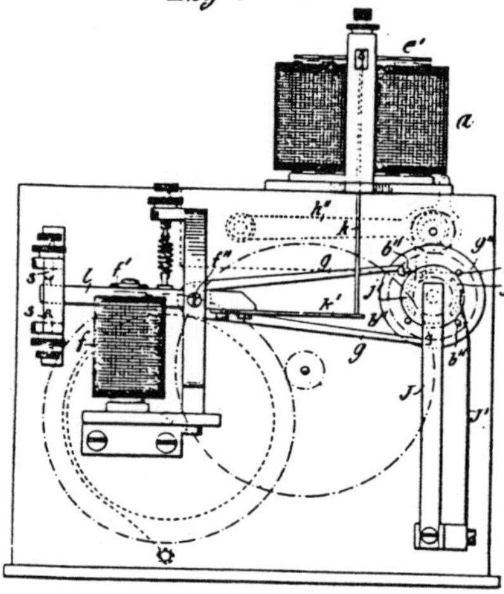

Fig. 6

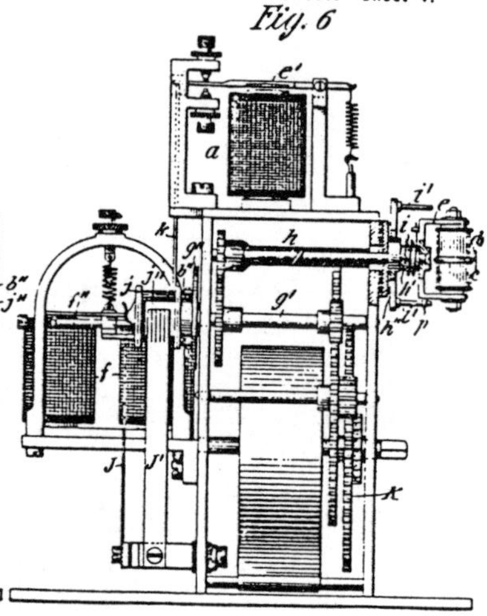

Fig. 4

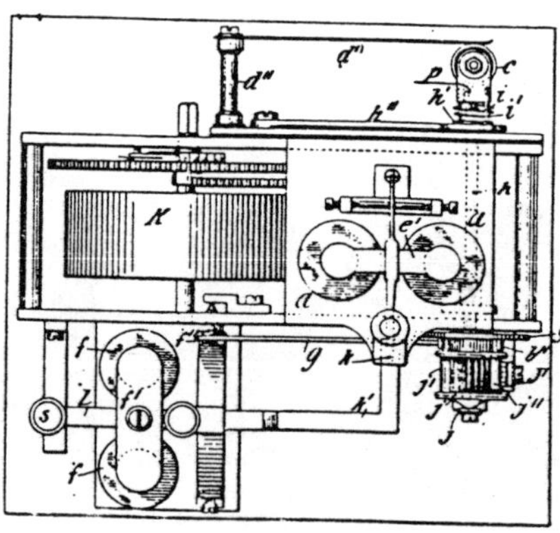

Fig. 8

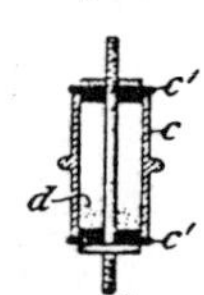

Fig. 7

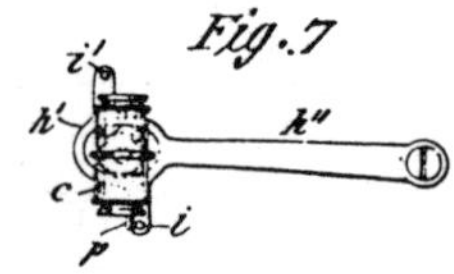

Witnesses:
Raphaël Netter
George Scherff.

Inventor:
Nikola Tesla

No. 613,809. Patented Nov. 8, 1898.

N. TESLA.

METHOD OF AND APPARATUS FOR CONTROLLING MECHANISM OF MOVING VESSELS OR VEHICLES.

(No Model.) 5 Sheets—Sheet 5.

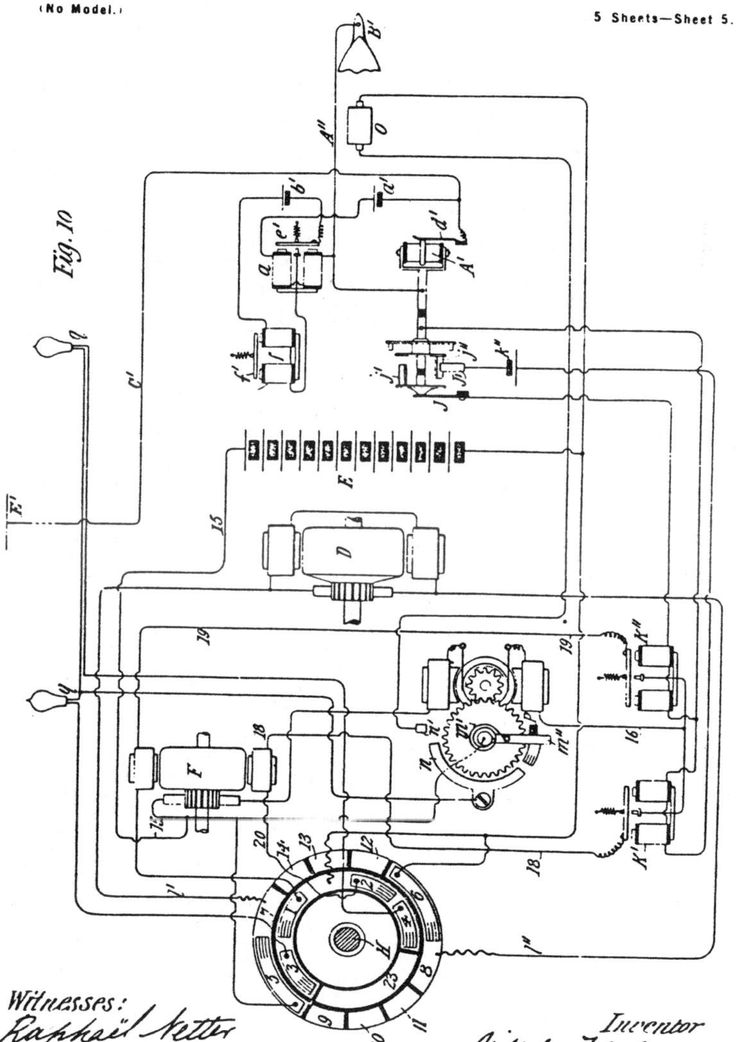

Fig. 10

Witnesses:
Raphaël Netter
M. Lawson Dyer

Inventor
Nikola Tesla
By Kerr, Curtis & Page
attys.

UNITED STATES PATENT OFFICE.

NIKOLA TESLA, OF NEW YORK, N. Y.

METHOD OF AND APPARATUS FOR CONTROLLING MECHANISM OF MOVING VESSELS OR VEHICLES.

SPECIFICATION forming part of Letters Patent No. 613,809, dated November 8, 1898.

Application filed July 1, 1898. Serial No. 684,934. (No model.)

To all whom it may concern:

Be it known that I, NIKOLA TESLA, a citizen of the United States, residing at New York, in the county and State of New York, have invented certain new and useful improvements in methods of and apparatus for controlling from a distance the operation of the propelling-engines, the steering apparatus, and other mechanism carried by moving bodies or floating vessels, of which the following is a specification, reference being had to the drawings accompanying and forming part of the same.

The problem for which the invention forming the subject of my present application affords a complete and practicable solution is that of controlling from a given point the operation of the propelling-engines, the steering apparatus, and other mechanism carried by a moving object, such as a boat or any floating vessel, whereby the movements and course of such body or vessel may be directed and controlled from a distance and any device carried by the same brought into action at any desired time. So far as I am aware the only attempts to solve this problem which have heretofore met with any measure of success have been made in connection with a certain class of vessels the machinery of which was governed by electric currents conveyed to the controlling apparatus through a flexible conductor; but this system is subject to such obvious limitations as are imposed by the length, weight, and strength of the conductor which can be practically used, by the difficulty of maintaining with safety a high speed of the vessel or changing the direction of movement of the same with the desired rapidity, by the necessity for effecting the control from a point which is practically fixed, and by many well-understood drawbacks inseparably connected with such a system. The plan which I have perfected involves none of these objections, for I am enabled by the use of my invention to employ any means of propulsion, to impart to the moving body or vessel the highest possible speed, to control the operation of its machinery and to direct its movements from either a fixed point or from a body moving and changing its direction however rapidly, and to maintain this control over great distances without any artificial connections between the vessel and the apparatus governing its movements and without such restrictions as these must necessarily impose.

In a broad sense, then, my invention differs from all of those systems which provide for the control of the mechanism carried by a moving object and governing its motion in that I require no intermediate wires, cables, or other form of electrical or mechanical connection with the object save the natural media in space. I accomplish, nevertheless, similar results and in a much more practicable manner by producing waves, impulses, or radiations which are received through the earth, water, or atmosphere by suitable apparatus on the moving body and cause the desired actions so long as the body remains within the active region or effective range of such currents, waves, impulses, or radiations.

The many and difficult requirements of the object here contemplated, involving peculiar means for transmitting to a considerable distance an influence capable of causing in a positive and reliable manner these actions, necessitated the designing of devices and apparatus of a novel kind in order to utilize to the best advantage various facts or results, which, either through my own investigations or those of others, have been rendered practically available.

As to that part of my invention which involves the production of suitable waves or variations and the conveying of the same to a remote receiving apparatus capable of being operated or controlled by their influence, it may be carried out in various ways, which are at the present time more or less understood. For example, I may pass through a conducting-path, preferably inclosing a large area, a rapidly-varying current and by electromagnetic induction of the same affect a circuit carried by the moving body. In this case the action at a given distance will be the stronger the larger the area inclosed by the conductor and the greater the rate of change of the current. If the latter were generated in the ordinary ways, the rate of change, and consequently the distance at which the action would be practically available for the present purpose, would be very small; but by adopting such means as I have devised—that is, either by passing through the conducting-path currents of a specially-designed high-frequency alternator or, better still, those of a strongly-charged condenser—a very high rate of change may be obtained and the effective range of the influence thus extended over a vast area, and by carefully adjusting the circuit on the moving body so as to be in exact electromagnetic synchronism with the primary disturbances this influence may be utilized at great distances.

Another way to carry out my invention is to direct the currents or discharges of a high-frequency machine or condenser through a circuit one terminal of which is connected directly or inductively with the ground and the other to a body, preferably of large surface and at an elevation. In this case if the circuit on the moving body be similarly arranged or connected differences of potential on the terminals of the circuit either by conduction or electrostatic induction are produced and the same object is attained. Again, to secure the best action the receiving-circuit should be adjusted so as to be in electromagnetic synchronism with the primary source, as before; but in this instance it will be understood by those skilled in the art that if the number of vibrations per unit of time be the same the circuit should now have a length of conductor only one-half of that used in the former case.

Still another way is to pass the currents simply through the ground by connecting both the terminals of the source of high-frequency currents to earth at different and remote points and to utilize the currents spreading through the ground for affecting a receiving-circuit properly placed and adjusted. Again, in this instance if only one of the terminals of the receiving-circuit be connected to the ground, the other terminal being insulated, the adjustment as to synchronism with the source will require that under otherwise equal conditions the length of wire be half of that which would be used if both the terminals be connected or, generally, if the circuit be in the form of a closed loop or coil. Obviously also in the latter case the relative position of the receiving and transmitting circuits is of importance, whereas if the circuit be of the former kind—that is, open—the relative position of the circuits is, as a rule, of little or no consequence.

Finally, I may avail myself, in carrying out my invention, of electrical oscillations which do not follow any particular conducting-path, but propagate in straight lines through space, of rays, waves, pulses, or disturbances of any kind capable of bringing the mechanism of the moving body into action from a distance and at the will of the operator by their effect upon suitable controlling devices.

In the following detailed description I shall confine myself to an explanation of that method and apparatus only which I have found to be the most practical and effectual; but obviously my invention in its broad features is not limited to the special mode and appliances which I have devised and shall here describe.

In any event—that is to say, whichever of the above or similar plans I may adopt—and particularly when the influence exerted from a distance upon the receiving-circuit be too small to directly and reliably affect and actuate the controlling apparatus I employ auxiliary sensitive relays or, generally speaking, means capable of being brought into action by the feeblest influences in order to effect the control of the movements of the distant body with the least possible expenditure of energy and at the greatest practicable distance, thus extending the range and usefulness of my invention.

A great variety of electrical and other devices more or less suitable for the purpose of detecting and utilizing feeble actions are now well known to scientific men and artisans and need not be all enumerated here. Confining myself merely to the electrical as the most practicable of such means and referring only to those which, while not the most sensitive, are perhaps more readily available from the more general knowledge which exists regarding them, I may state that a contrivance may be used which has long been known and used as a lightning-arrester in connection with telephone-switchboards for operating annunciators and like devices, comprising a battery the poles of which are connected to two conducting-terminals separated by a minute thickness of dielectric. The electromotive force of the battery should be such as to strain the thin dielectric layer very nearly to the point of breaking down in order to increase the sensitiveness. When an electrical disturbance reaches a circuit so arranged and adjusted, additional strain is put upon the insulating-film, which gives way and allows the passage of a current which can be utilized to operate any form of circuit-controlling apparatus.

Again, another contrivance capable of being utilized in detecting feeble electrical effects consists of two conducting plates or terminals which have, preferably, wires of some length attached to them and are bridged by a mass of minute particles of metal or other conducting material. Normally these particles lying loose do not connect the metal plates; but under the influence of an electrical disturbance produced at a distance, evidently owing to electrostatic attraction, they are pressed firmly against each other, thus establishing a good electrical connection between the two terminals. This change of state may be made use of in a number of ways for the above purpose.

Still another modified device, which may be said to embody the features of both the former, is obtained by connecting the two conducting plates or terminals above referred to permanently with the poles of a battery which should be of very constant electromotive force. In this arrangement a distant electrical disturbance produces a twofold effect on the conducting particles and insulating-films between them. The former are brought nearer to each other in consequence of the sudden increase of electrostatic attraction, and the latter, owing to this, as well as by being reduced in thickness or in number, are subjected to a much greater strain, which they are unable to withstand.

It will be obviously noted from the preceding that whichever of these or similar contrivances be used the sensitiveness and, what is often still more important, the reliability of operation is very materially increased by a close adjustment of the periods of vibration of the transmitting and receiving circuits, and, although such adjustment is in many cases unnecessary for the successful carrying out of my invention, I nevertheless make it a rule to bestow upon this feature the greatest possible care, not only because of the above-mentioned advantages, which are secured by the observance of the most favorable conditions in this respect, but also and chiefly with the object of preventing the receiving-circuit from being affected by waves or disturbances emanating from sources not under the control of the operator. The narrower the range of vibrations which are still capable of perceptibly affecting the receiving-circuit the safer will the latter be against extraneous disturbances. To secure the best result, it is necessary, as is well known to experts, to construct the receiving-circuit or that part of the same in which the vibration chiefly occurs so that it will have the highest possible self-induction and at the same time the least possible resistance. In this manner I have demonstrated the practicability of providing a great number of such receiving-circuits—fifty or a hundred, or more—each of which may be called up or brought into action whenever desired without the others being interfered with. This result makes it possible for one operator to direct simultaneously the movements of a number of bodies as well as to control the action of a number of devices located on the same body, each of which may have a distinct duty to fulfil. In the following description, however, I shall show a still further development in this direction—namely, how, by making use of merely one receiving-circuit, a great variety of devices may be actuated and any number of different functions performed at the will and command of the distant operator.

It should be stated in advance in regard to the sensitive devices above mentioned, which may be broadly considered as belonging to one class, inasmuch as the operation of all of them involves the breaking down of a minute thickness of highly-strained dielectric, that it is necessary to make some provision for automatically restoring to the dielectric its original unimpaired insulating qualities in order to enable the device to be used in successive operations. This is usually accomplished by a gentle tapping or vibration of the electrodes or particles or continuous rotation of the same; but in long experience with many forms of these devices I have found that such procedures, while suitable in simple and comparatively unimportant operations, as ordinary signaling, when it is merely required that the succeeding effects produced in the receiving-circuit should differ in regard to their relative duration only, in which case it is of little or no consequence if some of the individual effects be altered or incomplete or even entirely missed, do not yield satisfactory results in many instances, when it may be very important that the effects produced should all be exactly such as desired and that none should fail. To illustrate, let it be supposed that an official directing the movements of a vessel in the manner described should find it necessary to bring into action a special device on the latter or to perform a particular operation, perhaps of vital moment, at an instant's notice and possibly when, by design or accident, the vessel itself or any mark indicating its presence is hidden from his view. In this instance a failure or defective action of any part of the apparatus might have disastrous consequences and such cases in which the sure and timely working of the machinery is of paramount importance may often present themselves in practice, and this consideration has impressed me with the necessity of doing away with the defects in the present devices and procedures and of producing an apparatus which while being sensitive will also be most reliable and positive in its action. In the arrangement hereinafter described these defects are overcome in a most satisfactory manner, enabling thousands of successive operations, in all respects alike, being performed by the controlling apparatus without a single irregularity or miss being recorded. For a better understanding of these and other details of the invention as I now carry them out I would refer to the accompanying drawings, in which—

Figure 1 is a plan view of a vessel and mechanism within the same. Fig. 2 is a longitudinal section of the same, showing the interior mechanism in side elevation. Fig. 3 is a plan view, partially diagrammatical, of the vessel, apparatus, and circuit connections of the same. Fig. 4 is a plan view, on an enlarged scale, of a portion of the controlling mechanism. Fig. 5 is an end view of the same. Fig. 6 shows the same mechanism in side elevation. Fig. 7 is a side view of a detail of the mechanism. Fig. 8 is a central sectional view, on a larger scale, of a sensitive device forming part of the receiving-circuit. Fig. 9 is a diagrammatic illustration of the system in its preferred form. Fig. 10 is a view of the various mechanisms employed, but on a larger scale, and leaving out or indicating conventionally certain parts of well-understood character.

Referring to Figs. 1 and 2, A designates any type of vessel or vehicle which is capable of being propelled and directed, such as a boat, a balloon, or a carriage. It may be designed to carry in a suitable compartment B objects of any kind, according to the nature of the uses to which it is to be applied. The vessel—in this instance a boat—is provided with suitable propelling machinery, which is shown as comprising a screw-propeller C, secured to the shaft of an electromagnetic motor D, which derives its power from storage batteries E E E E. In addition to the propelling engine or motor the boat carries also a small steering-motor F, the shaft of which is extended beyond its bearings and provided with a worm which meshes with a toothed wheel G. This latter is fixed to a sleeve b, freely movable on a vertical rod H, and is rotated in one or the other direction, according to the direction of rotation of the motor F.

The sleeve b on rod H is in gear, through the cog-wheels H' and H'', with a spindle G, mounted in vertical bearings at the stem of the boat and carrying the rudder F'.

The apparatus by means of which the operation of both the propelling and steering mechanisms is controlled involves, primarily, a receiving-circuit, which for reasons before stated is preferably both adjusted and rendered sensitive to the influence of waves or impulses emanating from a remote source, the adjustment being so that the period of oscillation of the circuit is either the same as that of the source or a harmonic thereof.

The receiving-circuit proper (diagrammatically shown in Figs. 3 and 10) comprises a terminal E', conductor C', a sensitive device A', and a conductor A'', leading to the ground conveniently through a connection to the metal keel B' of the vessel. The terminal E' should present a large conducting-surface and should be supported as high as practicable on a standard D', which is shown as broken in Fig. 2; but such provisions are not always necessary. It is important to insulate very well the conductor C' in whatever manner it be supported.

The circuit or path just referred to forms also a part of a local circuit, which latter includes a relay-magnet a and a battery a', the electromotive force of which is, as before explained, so determined that although the dielectric layers in the sensitive device A' are subjected to a great strain, yet normally they withstand the strain and no appreciable current flows through the local circuit; but when an electrical disturbance reaches the circuit the dielectric films are broken down, the resistance of the device A' is suddenly and greatly diminished, and a current traverses the relay-magnet A.

The particular sensitive device employed is shown in general views and in detail in Figs. 4, 6, 7, and 8. It consists of a metal cylinder c, with insulating-heads c', through which passes a central metallic rod c''. A small quantity of grains d of conducting material, such as an oxidized metal, is placed in the cylinder. A metallic strip d', secured to an insulated post d'', bears against the side of the cylinder c, connecting it with the conductor C', forming one part of the circuit. The central rod c'' is connected to the frame of the instrument and so to the other part of the circuit through the forked metal arm e, the ends of which are fastened with two nuts to the projecting ends of the rod, by which means the cylinder c is supported.

In order to interrupt the flow of battery-current which is started through the action of the sensitive device A', special means are provided, which are as follows: The armature e' of the magnet a, when attracted by the latter, closes a circuit containing a battery b' and magnet f. The armature-lever f' of this magnet is fixed to a rock-shaft f'', to which is secured an anchor-escapement g, which controls the movements of a spindle g', driven by a clock-train K. The spindle g' has fixed to it a disk g'' with four pins b'', so that for each oscillation of the escapement g the spindle g' is turned through one-quarter of a revolution. One of the spindles in the clock-train, as h, is geared so as to make one-half of a revolution for each quarter-revolution of spindle g'. The end of the former spindle extends through the side of the frame and carries an eccentric cylinder h', which passes through a slot in a lever h'', pivoted to the side of the frame. The forked arm e, which supports the cylinder c, is pivoted to the end of eccentric h', and the eccentric and said arm are connected by a spiral spring l. Two pins i' i' extend out from the lever h'', and one of these is always in the path of a projection on arm e. They operate to prevent the turning of cylinder c with the spindle h and the eccentric. It will be evident that a half-revolution of the spindle h will wind up the spring i and at the same time raise or lower the lever h'', and these parts are so arranged that just before the half-revolution of the spindle is completed the pin i', in engagement with projection or stop-pin p, is withdrawn from its path, and the cylinder c, obeying the force of the spring i, is suddenly turned end for end, its motion being checked by the other pin i'. The adjustment relatively to armature f' of magnet f is furthermore so made that the pin i' is withdrawn at the moment when the armature has nearly reached its extreme position in its approach toward the magnet—that is, when the lever l, which carries the armature f', almost touches the lower one of the two stops s s, Fig. 5—which limits its motion in both directions.

The arrangement just described has been the result of long experimenting with the object of overcoming certain defects in devices of this kind, to which reference has been made before. These defects I have found to be due to many causes, as the unequal size, weight, and shape of the grains, the unequal pressure which results from this and from the manner in which the grains are usually agitated, the lack of uniformity in the conductivity of the surface of the particles owing to the varying thickness of the superficial oxidized layer, the varying condition of the gas or atmosphere in which the particles are immersed, and to certain deficiencies, well known to experts, of the transmitting apparatus as heretofore employed, which are in a large measure reduced by the use of my improved high-frequency coils. To do away with the defects in the sensitive device, I prepare the particles so that they will be in all respects as nearly alike as possible. They are manufactured by a special tool, insuring their equality in size, weight, and shape, and are then uniformly oxidized by placing them for a given time in an acid solution of predetermined strength. This secures equal conductivity of their surfaces and stops their further deterioration, thus preventing a change in the character of the gas in the space in which they are inclosed. I prefer not to rarefy the atmosphere within the sensitive device, as this has the effect of rendering the former less constant in regard to its dielectric properties, but merely secure an air-tight inclosure of the particles and rigorous absence of moisture, which is fatal to satisfactory working.

The normal position of the cylinder *c* is vertical, and when turned in the manner described the grains in it are simply shifted from one end to the other; but inasmuch as they always fall through the same space and are subjected to the same agitation they are brought after each operation of the relay to precisely the same electrical condition and offer the same resistance to the flow of the battery-current until another impulse from afar reaches the receiving-circuit.

The relay-magnet *a* should be of such character as to respond to a very weak current and yet be positive in its action. To insure the retraction of its armature *e'* after the current has been established through the magnet *f* and interrupted by the inversion of the sensitive device *c*, a light rod *k* is supported in guides on the frame in position to be lifted by an extension *k'* of the armature-lever *l* and to raise slightly the armature *e*. As a feeble current may normally flow through the sensitive device and the relay-magnet *a*, which would be sufficient to hold though not draw the armature down, it is well to observe this precaution.

The operation of the relay-magnet *a* and the consequent operation of the electromagnet *f*, as above described, are utilized to control the operation of the propelling-engine and the steering apparatus in the following manner: On the spindle *g'*, which carries the escapement-disk *g''*, Figs. 4 and 6, is a cylinder *j* of insulating material with a conducting plate or head at each end. From these two heads, respectively, contact plates or segments *j' j''* extend on diametrically opposite sides of the cylinder. The plate *j''* is in electrical connection with the frame of the instrument through the head from which it extends, while insulted strips or brushes J J' bear upon the free end or head of the cylinder and the periphery of the same, respectively. Three terminals are thus provided, one always in connection with plate *j'*, the other always in connection with the plate *j''*, and the third adapted to rest on the strips *j'* and *j''* in succession or upon the intermediate insulating-spaces, according to the position in which the commutator is brought by the clock-train and the anchor-escapement *g*.

K' K'', Figs. 1, 3, and 10, are two relay-magnets conveniently placed in the rear of the propelling-engine. One terminal of a battery *k''* is connected to one end of each of the relay-coils, the opposite terminal to the brush J', and the opposite ends of the relay-coils to the brush J and to the frame of the instrument, respectively. As a consequence of this arrangement either the relay K' or K'' will be energized as the brush J' bears upon the plate *j'* or *j''*, respectively, or both relays will be inactive while the brush J' bears upon an insulating-space between the plates *j'* and *j''*. While one relay, as K', is energized, its armature closes a circuit through the motor F, which is rotated in a direction to throw the rudder to port. On the other hand, when relay K'' is active another circuit through the motor F is closed, which reverses its direction of rotation and shifts the rudder to starboard. These circuits, however, are at the same time utilized for other purposes, and their course is, in part, through apparatus which I shall describe before tracing their course.

The fixed rod H carries an insulating disk or head L, Fig. 2, to the under side of which are secured six brushes, 1, 2, 3, 4, 5, and 6, Fig. 3. The sleeve *b*, which surrounds the rod and is turned by the steering-motor F, carries a disk L', upon the upper face of which are two concentric circles of conducting contact-plates. Brushes 1, 2, 3, and 4 bear upon the inner circle of contacts, while the brushes 5 and 6 bear upon the outer circle of contacts. The outer circle of contacts comprises two long plates 7 and 8 on opposite sides of the disk and a series of shorter plates 9, 10, 11, 12, 13, and 14 in the front and rear. Flexible conductors *l' l''* connect the plates 7 and 8 with the terminals of the propelling-motor D, and the poles of the main battery E are connected to the brushes 5 and 6, respectively, so that while the rudder is straight or turned up to a certain angle to either side the current is conveyed through the brushes 5 and 6 and segments 7 and 8 to the propelling-motor D. The steering-motor F is also driven by current taken from the main battery E in the following manner: A conductor 15 from one pole of the battery leads to one of the commutator-brushes, and from the other brush runs a conductor 16 to one of the contacts of each relay K' K''. When one of these relays, as K'', is active, it continues this circuit through a wire 19 through one field-coil or set of coils on the motor F and thence to the brush 1. In a similar manner when the other relay K' is active the circuit is continued from wire 18 through a wire 20, the second or reversing set of field-coils, and to brush 2.

Both brushes 1 and 2 at all times when the rudder is not turned more than about forty-five degrees to one side are in contact with a long conducting-plate 21, and one brush in any position of the rudder is always in contact with said plate, and the latter is connected by a flexible conductor 22 with the opposite pole of the main battery. Hence the motor F may always be caused to rotate in one direction whatever may be the position of the rudder, and may be caused to rotate in either direction whenever the position of the rudder is less than a predetermined angle, conveniently forty-five degrees from the center position. In order, however, to prevent the rudder from being turned too far in either direction, the isolated plate 23 is used. Any movement of the rudder beyond a predetermined limit brings this plate under one or the other of the brushes 1 2 and breaks the circuit of motor F, so that the rudder can be driven no farther in that direction, but, as will be understood, the apparatus is in condition to turn the rudder over to the other side. In like manner the circuit of the propelling-motor D is controlled through brushes 5 and 6 and the segments on the outer circle of contacts of head L. If the short segments on either side of the circle are insulated, the motor D will be stopped whenever one of the brushes 5 or 6 passes onto one of them from the larger segments 7 8.

It is important to add that on all contact-points where a break occurs provision should be made to overcome the sparking and prevent the oscillation of electrical charges in the circuits, as such sparks and oscillations may affect the sensitive device. It is this consideration chiefly which makes it advisable to use the two relays K' K'', which otherwise might be dispensed with. They should be also placed as far as practicable from the sensitive device in order to guard the latter against any action of strong varying currents.

In addition to the mechanism described the vessel may carry any other devices or apparatus as might be required for accomplishing any special object of more or less importance. By way of illustration a small motor *m* is shown, Figs. 1 and 3, which conveniently serves for a number of purposes. This motor is shown connected in series with the armature of the steering-motor F, so that whenever either one of the circuits of the latter is closed through relays K' K'' the motor *m* is likewise rotated, but in all cases in the same direction. Its rotation is opposed by a spring *m'*, so that in normal operation, owing to the fact that the circuits of motor F are closed but a short time, the lever *m''*, which is fastened to one of the wheels of clockwork M, with which the armature of the motor is geared, will move but a short distance and upon cessation of the current return to a stop P; but if the circuits of the motor F are closed and opened rapidly in succession, which operation leaves the rudder unaffected, then the lever *m''* is moved to a greater angle, coming in contact with a metal plate *n*, and finally, if desired, with a post *n'*. Upon the lever *m''* coming in contact with plate *n* the current of the main battery passes either through one or other or both of the lights supported on standards *q q*, according to the position of brushes 3 and 4 relatively to the insulating-segment 23; but since the head L, carrying the segments, is geared to the rudder the position of the latter is in a general way determined by observing the lights. Both of the lights may be colored, and by flashing them up whenever desired the operator may guide at night the vessel in its course. For such purposes also the standards *r r* are provided, which should be painted in lively colors, so as to be visible by day at great distances. By opening and closing the circuits of motor F a greater number of times, preferably determined beforehand, the lever *m''* is brought in contact with post *n'*, thus closing the circuit of the main battery through a device *o* and bringing the latter into action at the moment desired. By similar contrivances or such as will readily suggest themselves to mechanicians any number of different devices may be operated.

Referring now to Fig. 9, which illustrates diagrammatically the system as practiced when directing the movements of a boat, in this figure S designates any source of electrical disturbance or oscillations the generation of which is controlled by a suitable switch contained in box T. The handle of the switch is movable in one direction only and stops on four points *t t' u u'*, so that as the handle passes from stop to stop oscillations are produced by the source during a very short time interval. There are thus produced four disturbances during one revolution and the receiving-circuit is affected four times; but it will be understood from the foregoing description of the controlling devices on the vessel that the rudder will be moved twice, once to right and once to left. Now I preferably place the handle of the switch so that when it is arrested on points *t t'*—that is, to the right or left of the operator—he is reminded that the vessel is being deflected to the right or left from its course, by which means the control is facilitated. The normal positions of the handle are therefore at *u u'* when the rudder is not acted upon, and it remains on the points *u u'* only so long as necessary. Since, as before stated, the working of the apparatus is very sure, the operator is enabled to perform any such operations as provision is made for without even seeing the vessel.

The manner of using the apparatus and the operation of the several instrumentalities comprising the same is in detail as follows: Normally the plate L' is turned so that brush 2 rests upon the insulated segment 23 and brush 6 upon one of the insulated short segments in the rear of the circle. Under these conditions the rudder will be turned to starboard and the circuit of motor D interrupted between brushes 5 and 6. At the same time only one of the circuits of motor F—that controlled by relay K'—is capable of being closed, since brush 2, which connects with the other, is out of contact with the long segment 21. Assuming now that it is desired to start the vessel and direct it to a given point, the handle T is turned from its normal position on point u' to the point t on the switch-box. This sends out an electrical disturbance, which, passing through the receiving-circuit on the vessel, affects the sensitive device A' and starts the flow of current through the local circuit, including said device, the relay a, and the battery a'. This, as has been previously explained, turns the cylinder j and causes the brush J' to pass from insulation onto the contact j'. The battery k'' is thus closed through relay K'', and the latter closes that circuit of the motor F which, starting from plate 22, which is permanently connected with one pole of the main battery, is completed through the brush 1, the field of motor F, wire 19, the armature of relay K'', wire 16, the motor m, the brushes and commutator of motor F, and wire 15 to the opposite terminal of the battery E. Motor F is thus set in operation to shift the rudder to port; but the movement of plate L' which follows brings the brush 6 back onto segment 8 and closes the circuit of the propelling-motor which starts the vessel. The motor F is permitted to run until the rudder has been turned sufficiently to steer the vessel in the desired direction, when the handle T is turned to the point u. This produces another action of the relay a and brush J' is shifted onto insulation and both relays K' and K'' are inactive. The rudder remains in the position to which it has been shifted by the motor F. If it be then desired to shift it to starboard, or in the opposite direction to that in which it was last moved, the handle T is simply turned to point t' and allowed to remain there until the motor F, which is now operated by relay K', the circuit of which is closed by strip J' coming into contact with plate j'', has done its work. The movement of handle T to the next point throws out both relays K' and K'', and the next movement causes a shifting of the rudder to port, and so on. Suppose, however, that after the rudder has been set at any angle to its middle position it be desired to shift it still farther in the same direction. In such case the handle is moved quickly over two points, so that the circuit which would move the rudder in the opposite direction is closed for too short a time interval to produce an appreciable effect and is allowed to rest on the third point until the rudder is shifted to the desired position, when the handle is moved to the next point, which again throws out both relays K' and K''. It will be understood that if the handle be held for a sufficiently long time upon either point t or t' the motor F will simply turn the plate L' in one direction or the other until the circuits of motors D and F are broken. It is furthermore evident that one relay K' or K'' will always be operative to start the motor F.

As previously explained, the longest period of operation of which the motor F is capable under ordinary conditions of use does not permit the motor m to shift the arm m' into contact with the plate n; but if the handle T be turned with a certain rapidity a series of current impulses will be directed through motor m; but as these tend to rotate the motor F in opposite directions they do not sensibly affect the latter, but act to rotate the motor m against the force of the coiled spring.

The invention which I have described will prove useful in many ways. Vessels or vehicles of any suitable kind may be used, as life, despatch, or pilot boats or the like, or for carrying letters, packages, provisions, instruments, objects, or materials of any description, for establishing communication with inaccessible regions and exploring the conditions existing in the same, for killing or capturing whales or other animals of the sea, and for many other scientific, engineering, or commercial purposes; but the greatest value of my invention will result from its effect upon warfare and armaments, for by reason of its certain and unlimited destructiveness it will tend to bring about and maintain permanent peace among nations.

Having now described my invention, what I claim is—

1. The improvement in the art of controlling the movements and operation of a vessel or vehicle herein described, which consists in producing waves or disturbances which are conveyed to the vessel by the natural media, actuating thereby suitable apparatus on the vessel and effecting the control of the propelling-engine, the steering and other mechanism by the operation of the said apparatus, as set forth.

2. The improvement in the art of controlling the movements and operation of a vessel or vehicle, herein described, which consists in establishing a region of waves or disturbances, and actuating by their influence exerted at a distance the devices on such vessel or vehicle, which control the propelling, steering and other mechanism thereon, as set forth.

3. The improvement in the art of controlling the movements and operation of a vessel

or vehicle, herein described, which consists in establishing a region of electrical waves or disturbances, and actuating by their influence, exerted at a distance, the devices on said vessel or vehicle, which control the propelling, steering and other mechanism thereon, as set forth.

4. The improvement in the art of controlling the movements and operation of a vessel or vehicle, herein described, which consists in providing on the vessel a circuit controlling the propelling, steering and other mechanism, adjusting or rendering such circuit sensitive to waves or disturbances of a definite character, establishing a region of such waves or disturbances, and rendering by their means the controlling-circuit active or inactive, as set forth.

5. The combination with a source of electrical waves or disturbances of a moving vessel or vehicle, and mechanism thereon for propelling, steering or operating the same, and a controlling apparatus adapted to be actuated by the influence of the said waves or disturbances at a distance from the source, as set forth.

6. The combination with a source of electrical waves or disturbances of a moving vessel or vehicle, mechanism for propelling, steering or operating the same, a circuit and means therein for controlling said mechanism, and means for rendering said circuit active or inactive through the influence of the said waves or disturbances exerted at a distance from the source, as set forth.

7. The combination with a source of electrical waves or disturbances and means for starting and stopping the same, of a vessel or vehicle, propelling and steering mechanism carried thereby, a circuit containing or connected with means for controlling the operation of said mechanism and adjusted or rendered sensitive to the waves or disturbances of the source, as set forth.

8. The combination with a source of electrical waves or disturbances, and means for starting and stopping the operation of the same, of a vessel or vehicle, propelling and steering mechanism carried thereby, local circuits controlling said mechanisms, a circuit sensitive to the waves or disturbances of the source and means therein adapted to control the said local circuits, as and for the purpose set forth.

9. The sensitive device herein described comprising in construction a receptacle containing a material such as particles of oxidized metal forming a part of the circuit, and means for turning the same end for end when the material has been rendered active by the passage through it of an electric discharge, as set forth.

10. The sensitive device herein described, comprising in combination a receptacle containing a material such as particles of oxidized metal forming a part of an electric circuit, an electromagnet in said circuit, and devices controlled thereby for turning the receptacle end for end when said magnet is energized, as set forth.

11. The sensitive device herein described, comprising in combination a receptacle containing a material such as particles of oxidized metal forming part of an electric circuit, a motor for rotating the receptacle, an electromagnet in circuit with the material, and an escapement controlled by said magnet and adapted to permit a half-revolution of the receptacle when the said magnet is energized, as set forth.

12. The combination with a movable body or vehicle, of a propelling-motor, a steering-motor and electrical contacts carried by a moving portion of the steering mechanism, and adapted in certain positions of the latter to interrupt the circuit of the propelling-motor, a local circuit and means connected therewith for controlling the steering-motor, and a circuit controlling the local circuit and means for rendering said controlling-circuit sensitive to the influence of electric waves or disturbances exerted at a distance from their source, at set forth.

13. The combination with the steering-motor, a local circuit for directing current through the same in opposite directions, a controlling-circuit rendered sensitive to the influence of electric waves or disturbances exerted at a distance from their source, a motor in circuit with the steering-motor but adapted to run always in the same direction, and a local circuit or circuits controlled by said motor, as set forth.

NIKOLA TESLA.

Witnesses:
RAPHAËL NETTER,
GEORGE SCHERFF.

No. 645,576.

Patented Mar. 20, 1900.

N. TESLA.

SYSTEM OF TRANSMISSION OF ELECTRICAL ENERGY.

(Application filed Sept. 2, 1897.)

(No Model.)

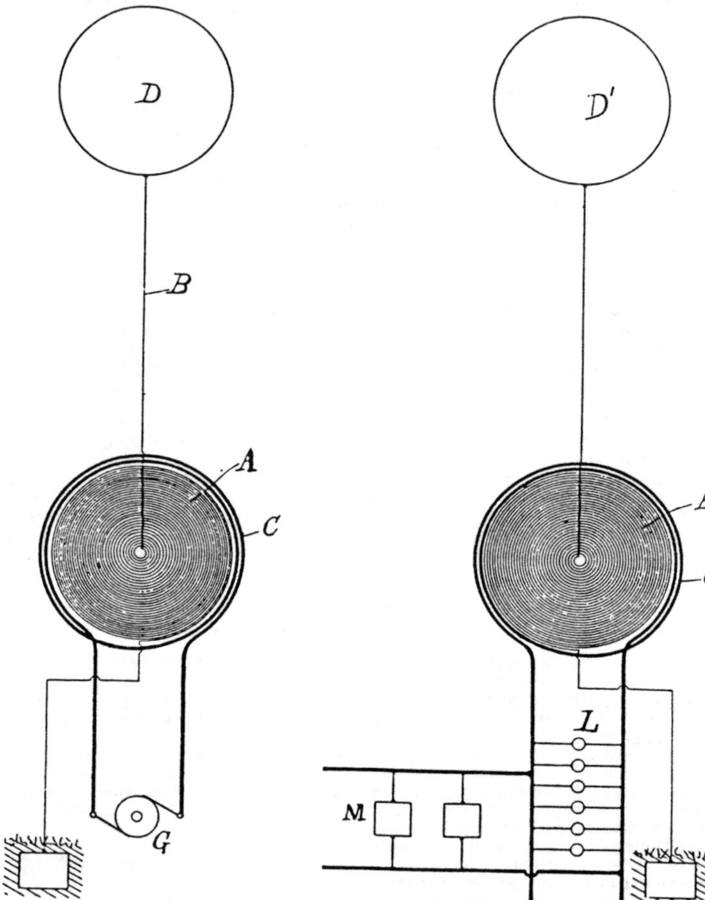

WITNESSES

Drury W. Cooper
M. Lawson Dyer

INVENTOR

Nikola Tesla

BY

Kerr, Curtis & Page

ATTORNEYS.

UNITED STATES PATENT OFFICE.

NIKOLA TESLA, OF NEW YORK, N. Y.

SYSTEM OF TRANSMISSION OF ELECTRICAL ENERGY.

SPECIFICATION forming part of Letters Patent No. 645,576, dated March 20, 1900.

Application filed September 2, 1897. Serial No. 650,343. (No model.)

To all whom it may concern:

Be it known that I, NIKOLA TESLA, a citizen of the United States, residing at New York, in the county and State of New York, have invented certain new and useful Improvements in Systems of Transmission of Electrical Energy, of which the following is a specification, reference being had to the drawing accompanying and forming a part of the same.

It has been well known heretofore that by rarefying the air inclosed in a vessel its insulating properties are impaired to such an extent that it becomes what may be considered as a true conductor, although one of admittedly very high resistance. The practical information in this regard has been derived from observations necessarily limited in their scope by the character of the apparatus or means heretofore known and the quality of the electrical effects producible thereby. Thus it has been shown by William Crookes in his classical researches, which have so far served as the chief source of knowledge of this subject, that all gases behave as excellent insulators until rarefied to a point corresponding to a barometric pressure of about seventy-five millimeters, and even at this very low pressure the discharge of a high-tension induction-coil passes through only a part of the attenuated gas in the form of a luminous thread or arc, a still further and considerable diminution of the pressure being required to render the entire mass of the gas inclosed in a vessel conducting. While this is true in every particular so long as electromotive or current impulses such as are obtainable with ordinary forms of apparatus are employed, I have found that neither the general behavior of the gases nor the known relations between electrical conductivity and barometric pressure are in conformity with these observations when impulses are used such as are producible by methods and apparatus devised by me and which have peculiar and hitherto unobserved properties and are of effective electromotive forces, measuring many hundred thousands or millions of volts. Through the continuous perfection of these methods and apparatus and the investigation of the actions of these current impulses I have been led to the discovery of certain highly-important and useful facts which have hitherto been unknown. Among these and bearing directly upon the subject of my present application are the following: First, that atmospheric or other gases, even under normal pressure, when they are known to behave as perfect insulators, are in a large measure deprived of their dielectric properties by being subjected to the influence of electromotive impulses of the character and magnitude I have referred to and assume conducting and other qualities which have been so far observed only in gases greatly attenuated or heated to a high temperature, and, second, that the conductivity imparted to the air or gases increases very rapidly both with the augmentation of the applied electrical pressure and with the degree of rarefaction, the law in this latter respect being, however, quite different from that heretofore established. In illustration of these facts a few observations, which I have made with apparatus devised for the purposes here contemplated, may be cited. For example, a conductor or terminal, to which impulses such as those here considered are supplied, but which is otherwise insulated in space and is remote from any conducting-bodies, is surrounded by a luminous flame-like brush or discharge often covering many hundreds or even as much as several thousands of square feet of surface, this striking phenomenon clearly attesting the high degree of conductivity which the atmosphere attains under the influence of the immense electrical stresses to which it is subjected. This influence is, however, not confined to that portion of the atmosphere which is discernible by the eye as luminous and which, as has been the case in some instances actually observed, may fill the space within a spherical or cylindrical envelop of a diameter of sixty feet or more, but reaches out to far remote regions, the insulating qualities of the air being, as I have ascertained, still sensibly impaired at a distance many hundred times that through which the luminous discharge projects from the terminal and in all probability much farther. The distance extends with the increase of the electromotive force of the impulses, with the diminution of the density of the atmosphere, with the elevation of the active terminal above the ground, and also, apparently, in a slight measure, with the degree of moisture contained in the air. I have likewise observed that this region of decidedly-noticeable influence continuously enlarges as time goes on, and the discharge is allowed to pass not unlike a conflagration which slowly spreads, this being possibly due to the gradual electrification or ionization of the air or to the formation of less insulating gaseous compounds. It is, furthermore, a fact that such discharges of extreme tensions, approximating those of lightning, manifest a marked tendency to pass upward away from the ground, which may be due to electrostatic repulsion, or possibly to slight heating and consequent rising of the electrified or ionized air. These latter observations make it appear probable that a discharge of this character allowed to escape into the atmosphere from a terminal maintained at a great height will gradually leak through and establish a good conducting-path to more elevated and better conducting air strata, a process which possibly takes place in silent lightning discharges frequently witnessed on hot and sultry days. It will be apparent to what an extent the conductivity imparted to the air is enhanced by the increase of the electromotive force of the impulses when it is stated that in some instances the area covered by the flame discharge mentioned was enlarged more than sixfold by an augmentation of the electrical pressure, amounting scarcely to more than fifty per cent. As to the influence of rarefaction upon the electric conductivity imparted to the gases it is noteworthy that, whereas the atmospheric or other gases begin ordinarily to manifest this quality at something like seventy-five millimeters barometric pressure with the impulses of excessive electromotive force to which I have referred, the conductivity, as already pointed out, begins even at normal pressure and continuously increases with the degree of tenuity of the gas, so that at, say, one hundred and thirty millimeters pressure, when the gases are known to be still nearly perfect insulators for ordinary electromotive forces, they behave toward electromotive impulses of several millions of volts like excellent conductors, as though they were rarefied to a much higher degree. By the discovery of these facts and the perfection of means for producing in a safe, economical, and thoroughly-practicable manner current impulses of the character described it becomes possible to transmit through easily-accessible and only moderately-rarefied strata of the atmosphere electrical energy not merely in insignificant quantities, such as are suitable for the operation of delicate instruments and like purposes, but also in quantities suitable for industrial uses on a large scale up to practically any amount and, according to all the experimental evidence I have obtained, to any terrestrial distance. To conduce to a better understanding of this method of transmission of energy and to distinguish it clearly, both in its theoretical aspect and in its practical bearing, from other known modes of transmission, it is useful to state that all previous efforts made by myself and others for transmitting electrical energy to a distance without the use of metallic conductors, chiefly with the object of actuating sensitive receivers, have been based, in so far as the atmosphere is concerned, upon those qualities which it possesses by virtue of its being an excellent insulator, and all these attempts would have been obviously recognized as ineffective if not entirely futile in the presence of a conducting atmosphere or medium. The utilization of any conducting properties of the air for purposes of transmission of energy has been hitherto out of the question in the absence of apparatus suitable for meeting the many and difficult requirements, although it has long been known or surmised that atmospheric strata at great altitudes—say fifteen or more miles above sea-level—are, or should be, in a measure, conducting; but assuming even that the indispensable means should have been produced then still a difficulty, which in the present state of the mechanical arts must be considered as insuperable, would remain—namely, that of maintaining terminals at elevations of fifteen miles or more above the level of the sea. Through my discoveries before mentioned and the production of adequate means the necessity of maintaining terminals at such inaccessible altitudes is obviated and a practical method and system of transmission of energy through the natural media is afforded essentially different from all those available up to the present time and possessing, moreover, this important practical advantage, that whereas in all such methods or systems heretofore used or proposed but a minute fraction of the total energy expended by the generator or transmitter was recoverable in a distant receiving apparatus by my method and appliances it is possible to utilize by far the greater portion of the energy of the source and in any locality however remote from the same.

Expressed briefly, my present invention, based upon these discoveries, consists then in producing at one point an electrical pressure of such character and magnitude as to cause thereby a current to traverse elevated strata of the air between the point of generation and a distant point at which the energy is to be received and utilized.

In the accompanying drawing a general arrangement of apparatus is diagrammatically illustrated such as I contemplate employing in the carrying out of my invention on an industrial scale—as, for instance, for lighting distant cities or districts from places where cheap power is obtainable.

Referring to the drawing, A is a coil, generally of many turns and of a very large diameter, wound in spiral form either about a magnetic core or not, as may be found necessary. C is a second coil, formed of a conductor of much larger section and smaller length, wound around and in proximity to the coil A. In the transmitting apparatus the coil A constitutes the high-tension secondary and the coil C the primary of much lower tension of a transformer. In the circuit of the primary C is included a suitable source of current G. One terminal of the secondary A is at the center of the spiral coil, and from this terminal the current is led by a conductor B to a terminal D, preferably of large surface, formed or maintained by such means as a balloon at an elevation suitable for the purposes of transmission, as before described. The other terminal of the secondary A is connected to earth and, if desired, also to the primary in order that the latter may be at substantially the same potential as the adjacent portions of the secondary, thus insuring safety. At the receiving-station a transformer of similar construction is employed; but in this case the coil A', of relatively-thin wire, constitutes the primary and the coil C', of thick wire or cable, the secondary of the transformer. In the circuit of the latter are included lamps L, motors M, or other devices for utilizing the current. The elevated terminal D' is connected with the center of the coil A', and the other terminal of said coil is connected to earth and preferably, also, to the coil C' for the reasons above stated.

It will be observed that in coils of the character described the potential gradually increases with the number of turns toward the center, and the difference of potential between the adjacent turns being comparatively small a very high potential, impracticable with ordinary coils, may be successfully obtained. It will be, furthermore, noted that no matter to what an extent the coils may be modified in design and construction, owing to their general arrangement and manner of connection, as illustrated, those portions of the wire or apparatus which are highly charged will be out of reach, while those parts of the same which are liable to be approached, touched, or handled will be at or nearly the same potential as the adjacent portions of the ground, this insuring, both in the transmitting and receiving apparatus and regardless of the magnitude of the electrical pressure used, perfect personal safety, which is best evidenced by the fact that although such extreme pressures of many millions of volts have been for a number of years continuously experimented with no injury has been sustained neither by myself or any of my assistants.

The length of the thin-wire coil in each transformer should be approximately one-quarter of the wave length of the electric disturbance in the circuit, this estimate being based on the velocity of propagation of the disturbance through the coil itself and the circuit with which it is designed to be used. By way of illustration if the rate at which the current traverses the circuit, including the coil, be one hundred and eighty-five thousand miles per second then a frequency of nine hundred and twenty-five per second would maintain nine hundred and twenty-five stationary waves in a circuit one hundred and eighty-five thousand miles long and each wave would be two hundred miles in length. For such a low frequency, to which I shall resort only when it is indispensable to operate motors of the ordinary kind under the conditions above assumed, I would use a secondary of fifty miles in length. By such an adjustment or porportioning of the length of wire in the secondary coil or coils the points of highest potential are made to coincide with the elevated terminals D D', and it should be understood that whatever length be given to the wires this condition should be complied with in order to attain the best results.

As the main requirement in carrying out my invention is to produce currents of an excessively-high potential, this object will be facilitated by using a primary current of very considerable frequency, since the electromotive force obtainable with a given length of conductor is proportionate to the frequency; but the frequency of the current is in a large measure arbitrary, for if the potential be sufficiently high and if the terminals of the coils be maintained at the proper altitudes the action described will take place, and a current will be transmitted through the elevated air strata, which will encounter little and possibly even less resistance than if conveyed through a copper wire of a practicable size. Accordingly the construction of the apparatus may be in many details greatly varied; but in order to enable any person skilled in the mechanical and electrical arts to utilize to advantage in the practical applications of my system the experience I have so far gained the following particulars of a model plant which has been long in use and which was constructed for the purpose of obtaining further data to be used in the carrying out of my invention on a large scale are given. The transmitting apparatus was in this case one of my electrical oscillators, which are transformers of a special type, now well known and characterized by the passage of oscillatory discharges of a condenser through the primary. The source G, forming one of the elements of the transmitter, was a condenser of a capacity of about four one-hundredths of a microfarad and was charged from a generator of alternating currents of fifty thousand volts pressure and discharged by means of a mechanically-operated break five thousand times per second through the primary C. The latter consisted of a single turn of stout stranded cable of inappreciable resistance and of an inductance of about eight thousand centimeters, the diameter of the loop being very nearly two hundred and forty-four centimeters. The total inductance of the primary circuit was approximately ten thousand centimeters, so that the primary circuit vibrated generally according to adjustment, from two hundred and thirty thousand to two hundred and fifty thousand times per second. The high-tension coil A in the form of a flat spiral was composed of fifty turns of heavily-insulated cable No. 8 wound in one single layer, the turns beginning close to the primary loop and ending near its center. The outer end of the secondary or high-tension coil A was connected to the ground, as illustrated, while the free end was led to a terminal placed in the rarefied air stratum through which the energy was to be transmitted, which was contained in an insulating-tube of a length of fifty feet or more, within which a barometric pressure varying from about one hundred and twenty to one hundred and fifty millimeters was maintained by means of a mechanical suction-pump. The receiving-transformer was similarly proportioned, the ratio of conversion being the reciprocal of that of the transmitter, and the primary high-tension coil A' was connected, as illustrated, with the end near the low-tension coil C' to the ground and with the free end to a wire or plate likewise placed in the rarefied air stratum and at the distance named from the transmitting-terminal. The primary and secondary circuits in the transmitting apparatus being carefully synchronized, an electromotive force from two to four million volts and more was obtainable at the terminals of the secondary coil A, the discharge passing freely through the attenuated air stratum maintained at the above barometric pressures, and it was easy under these conditions to transmit with fair economy considerable amounts of energy, such as are of industrial moment, to the receiving apparatus for supplying from the secondary coil C' lamps L or kindred devices. The results were particularly satisfactory when the primary coil or system A', with its secondary C', was carefully adjusted, so as to vibrate in synchronism with the transmitting coil or system A C. I have, however, found no difficulty in producing with apparatus of substantially the same design and construction electromotive forces exceeding three or four times those before mentioned and have ascertained that by their means current impulses can be transmitted through much-denser air strata. By the use of these I have also found it practicable to transmit notable amounts of energy through air strata not in direct contact with the transmitting and receiving terminals, but remote from them, the action of the impulses, in rendering conducting air of a density at which it normally behaves as an insulator, extending, as before remarked, to a considerable distance. The high electromotive force obtained at the terminals of coil or conductor A was, as will be seen, in the preceding instance, not so much due to a large ratio of transformation as to the joint effect of the capacities and inductances in the synchronized circuits, which effect is enhanced by a high frequency, and it will be obviously understood that if the latter be reduced a greater ratio of transformation should be resorted to, especially in cases in which it may be deemed of advantage to suppress as much as possible, and particularly in the transmitting-coil A, the rise of pressure due to the above effect and to obtain the necessary electromotive force solely by a large transformation ratio.

While electromotive forces such as are produced by the apparatus just described may be sufficient for many purposes to which my system will or may be applied, I wish to state that I contemplate using in an industrial undertaking of this kind forces greatly in excess of these, and with my present knowledge and experience in this novel field I would estimate them to range from twenty to fifty million volts and possibly more. By the use of these much greater forces larger amounts of energy may be conveyed through the atmosphere to remote places or regions, and the distance of transmission may be thus extended practically without limit.

As to the elevation of the terminals D D' it is obvious that it will be determined by a number of things, as by the amount and quality of the work to be performed, by the local density and other conditions of the atmosphere, by the character of the surrounding country, and such considerations as may present themselves in individual instances. Thus if there be high mountains in the vicinity the terminals should be at a greater height, and generally they should always be, if practicable, at altitudes much greater than those of the highest objects near them in order to avoid as much as possible the loss by leakage. In some cases when small amounts of energy are required the high elevation of the terminals, and more particularly of the receiving-terminal D', may not be necessary, since, especially when the frequency of the currents is very high, a sufficient amount of energy may be collected at that terminal by electrostatic induction from the upper air strata, which are rendered conducting by the active terminal of the transmitter or through which the currents from the same are conveyed.

With reference to the facts which have been pointed out above it will be seen that the altitudes required for the transmission of considerable amounts of electrical energy in accordance with this method are such as are easily accessible and at which terminals can be safely maintained, as by the aid of captive balloons supplied continuously with gas from reservoirs and held in position securely by steel wires or by any other means, devices, or expedients, such as may be contrived and perfected by ingenious and skilled engineers. From my experiments and observations I conclude that with electromotive impulses not greatly exceeding fifteen or twenty million volts the energy of many thousands of horse-power may be transmitted over vast distances, measured by many hundreds and even thousands of miles, with terminals not more than thirty to thirty-five thousand feet above the level of the sea, and even this comparatively-small elevation will be required chiefly for reasons of economy, and, if desired, it may be considerably reduced, since by such means as have been described practically any potential that is desired may be obtained, the currents through the air strata may be rendered very small, whereby the loss in the transmission may be reduced.

It will be understood that the transmitting as well as the receiving coils, transformers, or other apparatus may be in some cases movable—as, for example, when they are carried by vessels floating in the air or by ships at sea. In such a case, or generally, the connection of one of the terminals of the high-tension coil or coils to the ground may not be permanent, but may be intermittently or inductively established, and any such or similar modifications I shall consider as within the scope of my invention.

While the description here given contemplates chiefly a method and system of energy transmission to a distance through the natural media for industrial purposes, the principles which I have herein disclosed and the apparatus which I have shown will obviously have many other valuable uses—as, for instance, when it is desired to transmit intelligible messages to great distances, or to illuminate upper strata of the air, or to produce, designedly, any useful changes in the condition of the atmosphere, or to manufacture from the gases of the same products, as nitric acid, fertilizing compounds, or the like, by the action of such current impulses, for all of which and for many other valuable purposes they are eminently suitable, and I do not wish to limit myself in this respect. Obviously, also, certain features of my invention here disclosed will be useful as disconnected from the method itself—as, for example, in other systems of energy transmission, for whatever purpose they may be intended, the transmitting and receiving transformers arranged and connected as illustrated, the feature of a transmitting and receiving coil or conductor, both connected to the ground and to an elevated terminal and adjusted so as to vibrate in synchronism, the proportioning of such conductors or coils, as above specified, the feature of a receiving-transformer with its primary connected to earth and to an elevated terminal and having the operative devices in its secondary, and other features or particulars, such as have been described in this specification or will readily suggest themselves by a perusal of the same.

I do not claim in this application a transformer for developing or converting currents of high potential in the form herewith shown and described and with the two coils connected together, as and for the purpose set forth, having made these improvements the subject of a patent granted to me November 2, 1897, No. 593,138, nor do I claim herein the apparatus employed in carrying out the method of this application when such apparatus is specially constructed and arranged for securing the particular object sought in the present invention, as these last-named features are made the subject of an application filed as a division of this application on February 19, 1900, Serial No. 5,780.

What I now claim is—

1. The method hereinbefore described of transmitting electrical energy through the natural media, which consists in producing at a generating-station a very high electrical pressure, causing thereby a propagation or flow of electrical energy, by conduction, through the earth and the air strata, and collecting or receiving at a distant point the electrical energy so propagated or caused to flow.

2. The method hereinbefore described of transmitting electrical energy, which consists in producing at a generating-station a very high electrical pressure, conducting the current caused thereby to earth and to a terminal at an elevation at which the atmosphere serves as a conductor therefor, and collecting the current by a second elevated terminal at a distance from the first.

3. The method hereinbefore described of transmitting electrical energy through the natural media, which consists in producing between the earth and a generator-terminal elevated above the same, at a generating-station, a sufficiently-high electromotive force to render elevated air strata conducting, causing thereby a propagation or flow of electrical energy, by conduction, through the air strata, and collecting or receiving at a point distant from the generating-station the electrical energy so propagated or caused to flow.

4. The method hereinbefore described of transmitting electrical energy through the natural media, which consists in producing between the earth and a generator-terminal elevated above the same, at a generating-station, a sufficiently-high electromotive force to render the air strata at or near the elevated terminal conducting, causing thereby a propagation or flow of electrical energy, by conduction, through the air strata, and collecting or receiving at a point distant from the generating-station the electrical energy so propagated or caused to flow.

5. The method hereinbefore described of transmitting electrical energy through the natural media, which consists in producing between the earth and a generator-terminal elevated above the same, at a generating-station, electrical impulses of a sufficiently-high electromotive force to render elevated air strata conducting, causing thereby current impulses to pass, by conduction, through the air strata, and collecting or receiving at a point distant from the generating-station, the energy of the current impulses by means of a circuit synchronized with the impulses.

6. The method hereinbefore described of

transmitting electrical energy through the natural media, which consists in producing between the earth and a generator-terminal elevated above the same, at a generating-station, electrical impulses of a sufficiently-high electromotive force to render the air strata at or near the elevated terminal conducting, causing thereby current impulses to pass through the air strata, and collecting or receiving at a point distant from the generating-station the energy of the current impulses by means of a circuit synchronized with the impulses.

7. The method hereinbefore described of transmitting electrical energy through the natural media, which consists in producing between the earth and a generator-terminal elevated above the same, at a generating-station, electrical impulses of a wave length so related to the length of the generating circuit or conductor as to produce the maximum potential at the elevated terminal, and of sufficiently-high electromotive force to render elevated air strata conducting, causing thereby a propagation of electrical impulses through the air strata, and collecting or receiving at a point distant from the generating-station the energy of such impulses by means of a receiving-circuit having a length of conductor similarly related to the wave length of the impulses.

8. The method hereinbefore described of transmitting electrical energy through the natural media, which consists in producing between the earth and a generator-terminal elevated above the same, at a generating-station, a sufficiently-high electromotive force to render elevated air strata conducting, causing thereby a propagation or flow of electrical energy through the air strata, by conduction, collecting or receiving the energy so transmitted by means of a receiving-circuit at a point distant from the generating-station, using the receiving-circuit to energize a secondary circuit, and operating translating devices by means of the energy so obtained in the secondary circuit.

9. The method hereinbefore described of transmitting electrical energy through the natural media, which consists in generating current impulses of relatively-low electromotive force at a generating-station, utilizing such impulses to energize the primary of a transformer, generating by means of such primary circuit impulses in a secondary surrounding by the primary and connected to the earth and to an elevated terminal, of sufficiently-high electromotive force to render elevated air strata conducting, causing thereby impulses to be propagated through the air strata, collecting or receiving the energy of such impulses, at a point distant from the generating-station, by means of a receiving-circuit connected to the earth and to an elevated terminal, and utilizing the energy so received to energize a secondary circuit of low potential surrounding the receiving-circuit.

NIKOLA TESLA.

Witnesses:
M. LAWSON DYER,
G. W. MARTLING.

No. 685,953. Patented Nov. 5, 1901.

N. TESLA.

METHOD OF INTENSIFYING AND UTILIZING EFFECTS TRANSMITTED THROUGH NATURAL MEDIA.

(Application filed June 24, 1899. Renewed May 29, 1901.)

(No Model.)

Fig. 1.

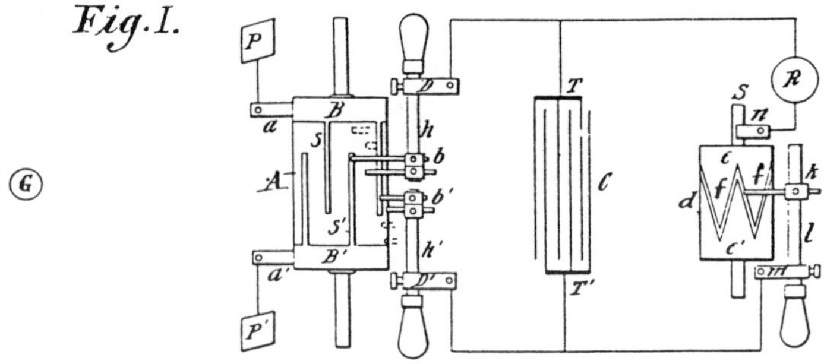

Fig. 2.

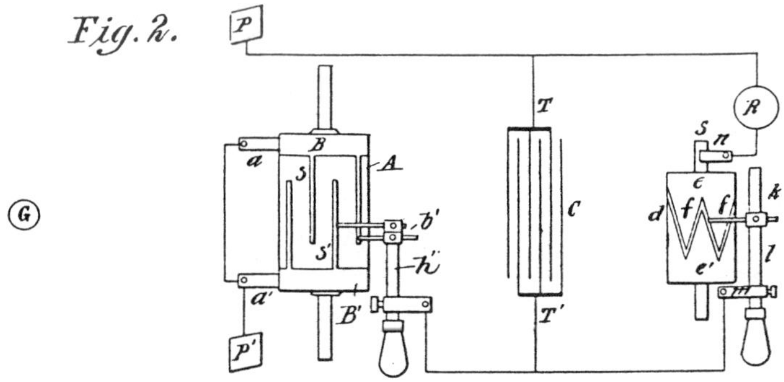

Witnesses:
G. B. Lewis.
Hellary C. Messimer.

Nikola Tesla, Inventor
by Kerr, Page & Cooper
Att'ys

UNITED STATES PATENT OFFICE.

NIKOLA TESLA, OF NEW YORK, N. Y.

METHOD OF INTENSIFYING AND UTILIZING EFFECTS TRANSMITTED THROUGH NATURAL MEDIA.

SPECIFICATION forming part of Letters Patent No. 685,953, dated November 5, 1901.

Application filed June 24, 1899. Renewed May 29, 1901. Serial No. 62,315. (No model.)

To all whom it may concern:

Be it known that I, NIKOLA TESLA, a citizen of the United States, residing at New York, in the county and State of New York, have invented a new and useful Improvement in Methods of Intensifying and Utilizing Effects Transmitted Through the Natural Media, of which the following is a specification, reference being had to the accompanying drawings, which form a part of the same.

The subject of my present invention is an improvement in the art of utilizing effects transmitted from a distance to a receiving device through the natural media; and it consists in a novel method by means of which results hitherto unattainable may be secured.

Several ways or methods of transmitting electrical disturbances through the natural media and utilizing them to operate distant receivers are now known and have been applied with more or less success for accomplishing a variety of useful results. One of these ways consists in producing by a suitable apparatus rays or radiations—that is, disturbances—which are propagated in straight lines through space, directing them upon a receiving or recording apparatus at a distance, and thereby bringing the latter into action. This method is the oldest and best known and has been brought particularly into prominence in recent years through the investigations of Heinrich Hertz. Another method consists in passing a current through a circuit, preferably one inclosing a very large area, inducing thereby in a similar circuit situated at a distance another current and affecting by the same in any convenient way a receiving device. Still another way, which has also been known for many years, is to pass in any suitable manner a current through a portion of the ground, as by connecting to two points of the same, preferably at a considerable distance from each other, the two terminals of a generator and to energize by a part of the current diffused through the earth a distant circuit which is similarly arranged and grounded at two points widely apart and which is made to act upon a sensitive receiver. These various methods have their limitations, one especially, which is common to all, being that the receiving circuit or instrument must be maintained in a definite position with respect to the transmitting apparatus, which often imposes great disadvantages upon the use of the apparatus.

In several applications filed by me and patents granted to me I have disclosed other methods of accomplishing results of this nature, which may be briefly described as follows: In one system the potential of a point or region of the earth is varied by imparting to it intermittent or alternating electrifications through one of the terminals of a suitable source of electrical disturbances which, to heighten the effect, has its other terminal connected to an insulated body, preferably of large surface and at an elevation. The electrifications communicated to the earth spread in all directions through the same, reaching a distant circuit which generally has its terminals arranged and connected similarly to those of the transmitting source and operates upon a highly-sensitive receiver. Another method is based upon the fact that the atmospheric air which behaves as an excellent insulator to currents generated by ordinary apparatus becomes a conductor under the influence of currents or impulses of enormously-high electromotive force which I have devised means for generating. By such means air strata, which are easily accessible, are rendered available for the production of many desired effects at distances, however great. This method, furthermore, allows advantage to be taken of many of those improvements which are practicable in the ordinary systems of transmission involving the use of a metallic conductor.

Obviously whatever method be employed it is desirable that the disturbances produced by the transmitting apparatus should be as powerful as possible, and by the use of certain forms of high-frequency apparatus which I have devised and which are now well known important practical advantages are in this respect secured. Furthermore, since in most cases the amount of energy conveyed to the distant circuit is but a minute fraction of the total energy emanating from the source it is necessary for the attainment of the best results that whatever the character of the receiver and the nature of the disturbances as much as possible of the energy conveyed should be made available for the operation of the receiver, and with this object in view I have heretofore among other means employed a receiving-circuit of high self-induction and very small resistance and of a period such as to vibrate in synchronism with the disturbances, whereby a number of separate impulses from the source were made to coöperate, thus magnifying the effect exerted upon and insuring the action of the receiving device. By these means decided advantages have been secured in many instances; but very often the improvement is either not applicable at all or, if so, the gain is very slight. Evidently when the source is one producing a continuous pressure or delivering impulses of long duration it is impracticable to magnify the effects in this manner and when, on the other hand, it is one furnishing short impulses of extreme rapidity of succession the advantage obtained in this way is insignificant, owing to the radiation and the unavoidable frictional waste in the receiving-circuit These losses reduce greatly both the intensity and the number of the coöperative impulses, and since the initial intensity of each of these is necessarily limited only an insignificant amount of energy is thus made available for a single operation of the receiver. As this amount is consequently dependent on the energy conveyed to the receiver by one single impulse it is evidently necessary to employ either a very large and costly, and therefore objectionable, transmitter or else to resort to the equally objectionable use of a receiving device too delicate and too easily deranged. Furthermore, the energy obtained through the coöperation of the impulses is in the form of extremely rapid vibrations and, because of this, unsuitable for the operation of ordinary receivers, the more so as this form of energy imposes narrow restrictions in regard to the mode and time of its application to such devices.

To overcome these and other limitations and disadvantages which have heretofore existed in such systems of transmission of signals or intelligence is the main object of my present invention, which comprises a novel method of accomplishing these ends.

The method, briefly stated, consists in producing arbitrarily-varied or intermittent disturbances or effects, transmitting such disturbances or effects through the natural media to a distant receiving-station, utilizing energy derived from such disturbances or effects at the receiving-station to charge a condenser, and using the accumulated potential energy so obtained to operate a receiving device.

An apparatus by means of which this method may be practiced is illustrated in the drawings hereto annexed, in which—

Figure 1 is a diagrammatic illustration of the apparatus, and Fig. 2 is a modified form or arrangement of the same.

In the practical application of my method I usually proceed as follows: At any two points in the transmitting medium between which there exists or may be obtained in any manner through the action of the disturbances or effects to be investigated or utilized a difference of electrical potential of any magnitude I arrange two plates or electrodes so that they may be oppositely charged through the agency of such effects or disturbances, and I connect these electrodes to the terminals of a highly-insulated condenser, generally of considerable capacity. To the condenser-terminals I also connect the receiver to be operated in series with a device of suitable construction, which performs the function of periodically discharging the condenser through the receiver at and during such intervals of time as may be best suitable for the purpose contemplated. This device may merely consist of two stationary electrodes separated by a feeble dielectric layer of minute thickness or it may comprise terminals one or more of which are movable and actuated by any suitable force and are adapted to be brought into and out of contact with each other in any convenient manner. It will now be readily seen that if the disturbances of whatever nature they may be cause definite amounts of electricity of the same sign to be conveyed to each of the plates or electrodes above mentioned, either continuously or at intervals of time which are sufficiently long, the condenser will be charged to a certain potential, and an adequate amount of energy being thus stored during the time determined by the device effecting the discharge of the condenser the receiver will be periodically operated by the electrical energy so accumulated; but very often the character of the impulses and the conditions of their use are such that without further provision not enough potential energy would be accumulated in the condenser to operate the receiving device. This is the case when, for example, each of the plates or terminals receives electricity of rapidly-changing sign or even when each receives electricity of the same sign, but only during periods which are short as compared with the intervals separating them. In such instances I resort to the use of a special device which I insert in the circuit between the plates and the condenser for the purpose of conveying to each of the terminals of the latter electrical charges of the proper quality and order of succession to enable the required amount of potential energy to be stored in the condenser.

There are a number of well-known devices, either without any moving parts or terminals or with elements reciprocated or rotated by the application of a suitable force, which offer a more ready passage to impulses of one sign or direction than to those of the other, or permit only impulses of one kind or order of succession to traverse a path, and any of these or similar devices capable of fulfilling the requirements may be used in carrying my invention into practice. One such device of familiar construction which will serve to convey a clear understanding of this part of my invention and enable a person skilled in the art to apply the same is illustrated in the annexed drawings. It consists of a cylinder A of insulating material, which is moved at a uniform rate of speed by clockwork or other suitable motive power and is provided with two metal rings B B', upon which bear brushes *a* and *a'*, which are connected, respectively, in the manner shown to the terminal plates P and P', above referred to. From the rings B B' extend narrow metallic segments *s* and *s'*, which by the rotation of the cylinder A are brought alternately into contact with double brushes *b* and *b'*, carried by and in contact with conducting-holders *h* and *h'*, which are adjustable longitudinally in the metallic supports D and D', as shown. The latter are connected to the terminals T and T' of a condenser C, and it should be understood that they are capable of angular displacement, as ordinary brush-supports. The object of using two brushes, as *b* and *b'*, in each of the holders *h* and *h'* is to vary at will the duration of the electric contact of the plates P and P' with the terminals T and T', to which is connected a receiving-circuit including a receiver R and a device *d* of the kind above referred to, which performs the duty of closing the receiving-circuit at predetermined intervals of time and discharging the stored energy through the receiver. In the present case this device consists of a cylinder *d*, made partly of conducting and partly of insulating material *e* and *e'*, respectively, which is rotated at the desired rate of speed by any suitable means. The conducting part *e* is in good electrical connection with the shaft S and is provided with tapering segments *f f*, upon which slides a brush *k*, supported on a conducting-rod *l*, capable of longitudinal adjustment in a metallic support *m*. Another brush *n* is arranged to bear upon the shaft S, and it will be seen that whenever one of the segments *f* comes in contact with the brush *k* the circuit, including the receiver R, is completed and the condenser discharged through the same. By an adjustment of the speed of rotation of the cylinder *d* and a displacement of the brush *k* along the cylinder the circuit may be made to open and close in as rapid succession and remain open or closed during such intervals of time as may be desired. The plates P and P', through which the electrifications are conveyed to the brushes *a* and *a'*, may be at a considerable distance from each other and both in the ground or both in the air, or one in the ground and the other in the air, preferably at some height, or they may be connected to conductors extending to some distance or to the terminals of any kind of apparatus supplying electrical energy which is obtained from the energy of the impulses or disturbances transmitted from a distance through the natural media.

In illustration of the operation of the devices described let it be assumed that alternating electrical impulses from a distant generator, as G, are transmitted through the earth and that it is desired to utilize those impulses in accordance with my method. This may be the case, for example, when such a generator is used for purposes of signaling in one of the ways before enumerated, as by having its terminals connected to two points of the earth distant from each other. In this case the plates P and P' are first connected to two properly-selected points of the earth. The speed of rotation of the cylinder A is varied until it is made to turn in synchronism with the alternate impulses of the generator, and, finally, the position of the brushes *b* and *b'* is adjusted by angular displacement, as usual, or in other ways, so that they are in contact with the segments *s* and *s'* during the periods when the impulses are at or near the maximum of their intensity. Only ordinary electrical skill and knowledge are required to make these adjustments, and a number of devices for effecting synchronous movement being well known, and it being the chief object of my present application to set forth a novel method of utilizing or applying a principle, a detailed description of such devices is not considered necessary. I may state, however, that for practical purposes in the present case it is only necessary to shift the brushes forward or back until the maximum effect is secured. The above requirements being fulfilled, electrical charges of the same sign will be conveyed to each of the condenser-terminals as the cylinder A is rotated, and with each fresh impulse the condenser will be charged to a higher potential. The speed of rotation of the cylinder *d* being adjustable at will, the energy of any number of separate impulses may thus be accumulated in potential form and discharged through the receiver R upon the brush *k* coming in contact with one of the segments *f*. It will be of course understood that the capacity of the condenser should be such as to allow the storing of a much greater amount of energy than is required for the ordinary operation of the receiver. Since by this method a relatively great amount of energy and in a suitable form may be made available for the operation of a receiver, the latter need not be very sensitive; but of course when the impulses are very feeble, as when coming from a great distance or when it is desired to operate a receiver very rapidly, then any of the well-known devices capable of responding to very feeble influences may be used in this connection.

If instead of the alternating impulses short impulses of the same direction are conveyed to the plates P and P', the apparatus described may still readily be used, and for this purpose it is merely necessary to shift the brushes *b* and *b'* into the position indicated by the dotted lines while maintaining the same conditions in regard to synchronism as before, so that the succeeding impulses will be permitted to pass into the condenser, but prevented from returning to the ground or transmitting medium during the intervals between them, owing to the interruption during such intervals of the connections leading from the condenser-terminals to the plates.

Another way of using the apparatus with impulses of the same direction is to take off one pair of brushes, as *b*, disconnect the plate P from brush *a* and join it directly to the terminal T of the condenser, and to connect brush *a* with brush *a'*. The apparatus thus modified would appear as shown in Fig. 2. Operated in this manner and assuming the speed of rotation of cylinder A to be the same, the apparatus will now be evidently adapted for a number of impulses per unit of time twice as great as in the preceding case. In all cases it is evidently important to adjust the duration of contact of segments *s* and *s'* with brushes *b b'* in the manner indicated.

When the method and apparatus I have described are used in connection with the transmission of signals or intelligence, it will of course be understood that the transmitter is operated in such a way as to produce disturbances or effects which are varied or intermitted in some arbitrary manner—for example, to produce longer and shorter successions of impulses corresponding to the dashes and dots of the Morse alphabet—and the receiving device will respond to and indicate these variations or intermittences, since the storage device will be charged and discharged a number of times corresponding to the duration of the successions of impulses received.

Obviously the special appliances used in carrying out my invention may be varied in many ways without departing from the spirit of the same.

It is to be observed that it is the function of the cylinder A, with its brushes and connections, to render the electrical impulses coming from the plates P and P' suitable for charging the condenser (assuming them to be unsuitable for this purpose in the form in which they are received) by rectifying them when they are originally alternating in direction or by selecting such parts of them as are suitable when all are not, and any other device performing this function will obviously answer the purpose. It is also evident that a device such as I have already referred to which offers a more ready passage to impulses of one sign or permits only impulses of the same sign to pass may also be used to perform this selective function in many cases when alternating impulses are received. When the impulses are long and all of the same direction, and even when they are alternating, but sufficiently long in duration and sustained in electromotive force, the brushes *b* and *b'* may be adjusted so as to bear on the parts B B' of the cylinder A, or the cylinder and its brushes may be omitted and the terminals of the condenser connected directly to the plates P and P'.

It will be seen that by the use of my invention results hitherto unattainable in utilizing disturbances or effects transmitted through natural media may be readily attained, since however great the distance of such transmission and however feeble or attenuated the impulses received enough energy may be accumulated from them by storing up the energy of succeeding impulses for a sufficient interval of time to render the sudden liberation of it highly effective in operating a receiver. In this way receivers of a variety of forms may be made to respond effectively to impulses too feeble to be detected or to be made to produce any sensible effect in any other way of which I am aware, a result of great value in various applications to practical use.

I do not claim herein an apparatus by means of which the above-described method is or may be practiced either in the special form herein shown or in other forms which are possible, having made claims to such apparatus in another application, Serial No. 729,812, filed September 8, 1899, as a division of the present case.

What I claim as my invention, and desire to secure by Letters Patent, is—

1. The method of transmitting and utilizing electrical energy herein described, which consists in producing arbitrarily varied or intermitted electrical disturbances or effects, transmitting the same to a distant receiving-station, charging, for succeeding and predetermined periods of time a condenser with energy derived from such effects or disturbances, and operating a receiving device by discharging at arbitrary intervals, the accumulated potential energy so obtained, as set forth.

2. The method of transmitting and utilizing electrical energy herein described, which consists in producing electrical disturbances or effects capable of being transmitted to a distance through the natural media, charging a condenser at a distant receiving-station with energy derived from such effects or disturbances, and using for periods of time, predetermined as to succession and duration, the potential energy so obtained to operate a receiving device.

3. The method of transmitting and utilizing electrical energy herein described, which consists in producing electrical disturbances or effects capable of being transmitted to a distance through the natural media, charging a condenser at a distant receiving-station for succeeding and predetermined periods of time, with energy derived from such effects or disturbances, and using for periods of time predetermined as to succession and duration, the accumulated energy so obtained to operate a receiving device.

4. The method hereinbefore described of producing arbitrarily varied or intermitted

electrical disturbances or effects, transmitting such disturbances or effects through the natural media to a distant receiving-station, storing in a condenser energy derived from a succession of such disturbances or effects for periods of time which correspond in succession to such effects or disturbances and are predetermined as to duration, and using the accumulated potential energy so obtained to operate a receiving device.

5. The method herein described of producing arbitrarily varied or intermitted electrical disturbances or effects, transmitting such disturbances or effects through the natural media to a distant receiving-station, establishing thereby a flow of electrical energy in a circuit at such station, charging a condenser with energy from such circuit, and using the accumulated potential energy so obtained to operate a receiving device.

6. The method herein described of producing arbitrarily varied or intermitted electrical disturbances or effects, transmitting such disturbances or effects through the natural media to a distant receiving-station, establishing thereby a flow of electrical energy in a circuit at such station, charging a condenser with electrical energy from such circuit, and discharging the accumulated potential energy so obtained into or through a receiving device at arbitrary intervals of time.

7. The method herein described of producing arbitrarily varied or intermitted electrical disturbances or effects, transmitting such disturbances or effects to a distant receiving-station, establishing thereby a flow of electrical energy in a circuit at such station, selecting or directing the impulses in said circuit so as to render them suitable for charging a condenser, charging a condenser with the impulses so selected or directed, and discharging the accumulated potential energy so obtained into, or through a receiving device.

8. The method herein described of producing arbitrarily varied or intermitted electrical disturbances or effects, transmitting such disturbances or effects through the natural media to a distant receiving-station, establishing thereby a flow of electrical energy in a circuit at such station, selecting or directing the impulses in said circuit so as to render them suitable for charging a condenser, charging a condenser with the impulses so selected or directed, and discharging the accumulated potential energy so obtained into, or through a receiving device at arbitrary intervals of time.

9. The method hereinbefore described of transmitting signals or intelligence, which consists in producing at the sending-station arbitrarily varied or intermitted disturbances or effects, transmitting such disturbances or effects through the natural media to a receiving-station, utilizing energy derived from such disturbances or effects at the receiving-station to charge a condenser and using the accumulated potential energy so obtained to operate a receiving device.

10. The method hereinbefore described of transmitting signals or intelligence through the natural media from a sending-station to a receiving-station, which consists in producing at the sending-station, arbitrarily varied or intermitted electrical effects or disturbances, transmitting the same through the natural media to the receiving-station, utilizing the energy derived from such disturbances or effects at the receiving-station to charge a condenser, and discharging the accumulated potential energy so obtained through a receiving device at arbitrary intervals of time.

11. The method hereinbefore described of transmitting signals or intelligence from a sending to a distant receiving station, which consists in producing at the former, arbitrarily varied or intermitted electrical disturbances or effects, transmitting the same to the receiving-station, charging by the energy derived from such disturbances or effects at the receiving-station a condenser, and using for periods of time predetermined as to succession and duration, the potential energy so obtained to operate a receiving device, as set forth.

NIKOLA TESLA.

Witnesses:
LEONARD E. CURTIS,
A. E. SKINNER.

No. 685,954 Patented Nov. 5, 1901.

N. TESLA.

METHOD OF UTILIZING EFFECTS TRANSMITTED THROUGH NATURAL MEDIA.

(Application filed Aug. 1, 1899. Renewed May 29, 1901.)

(No Model.) 2 Sheets—Sheet I.

Fig 1.

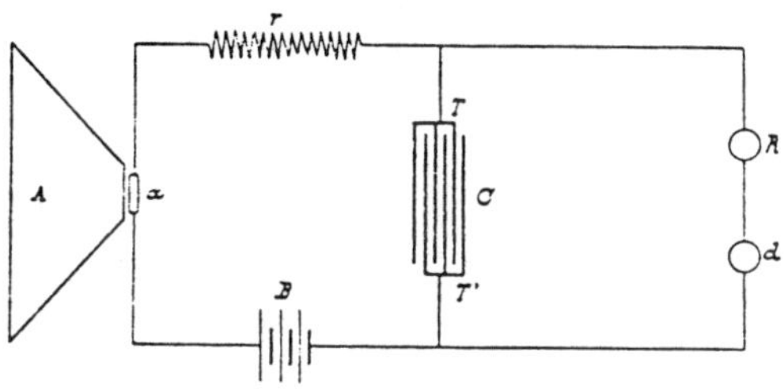

Fig 2.

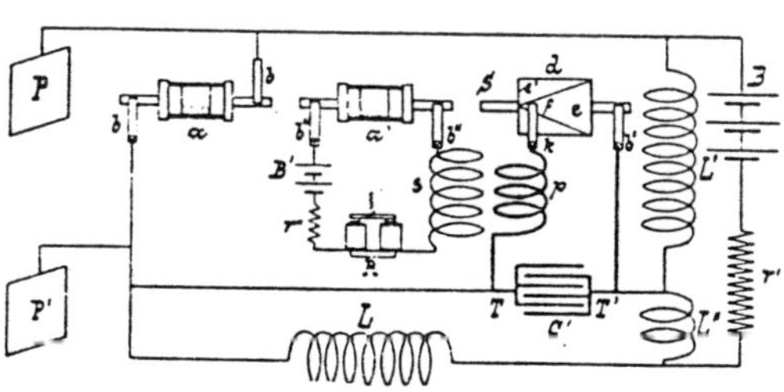

WITNESSES:
Benjamin Miller.
M. Lawson Dyer.

INVENTOR
Nikola Tesla
BY
Kerr, Page & Cooper
ATTORNEYS.

No. 685,954. Patented Nov. 5, 1901.

N. TESLA.

METHOD OF UTILIZING EFFECTS TRANSMITTED THROUGH NATURAL MEDIA.

(Application filed Aug. 1, 1899. Renewed May 29, 1901.)

(No Model.) 2 Sheets—Sheet 2.

Fig. 3

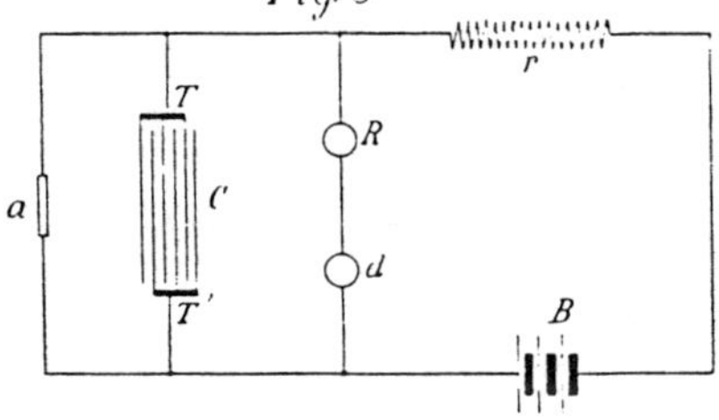

Fig. 4

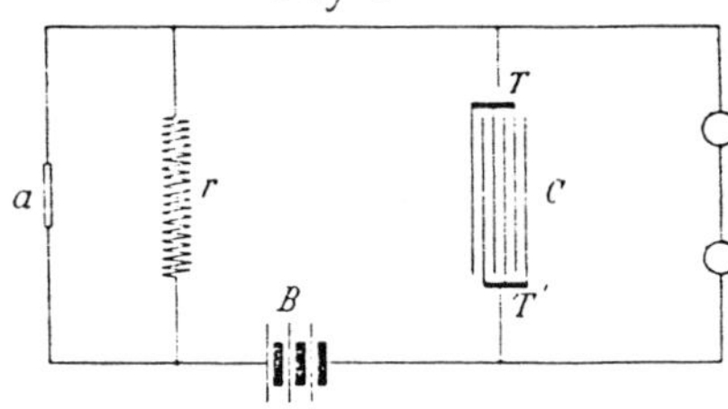

Fig. 5

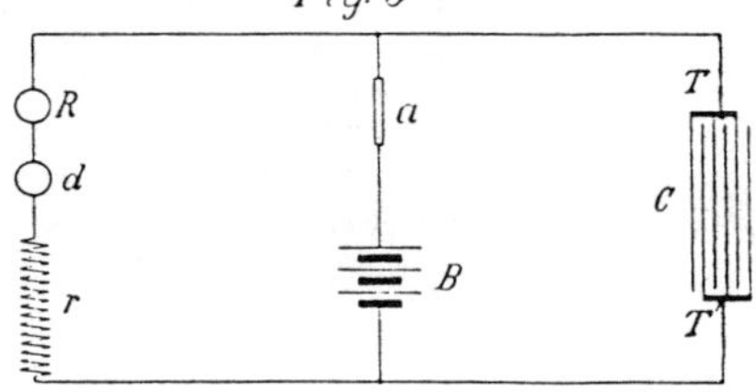

Witnesses:

Raphaël Netter

Benjamin Miller

Nikola Tesla Inventor

by Ken. Page & Cooper Att'ys

UNITED STATES PATENT OFFICE.

NIKOLA TESLA, OF NEW YORK, N. Y.

METHOD OF UTILIZING EFFECTS TRANSMITTED THROUGH NATURAL MEDIA.

SPECIFICATION forming part of Letters Patent No. 685,954, dated November 5, 1901.

Application filed August 1, 1899. Renewed May 29, 1901. Serial No. 62,318. (No model.)

To all whom it may concern:

Be it known that I, NIKOLA TESLA, a citizen of the United States, residing at New York city, in the county and State of New York, have invented a new and useful Improvement in Methods of Utilizing Effects Transmitted from a Distance to a Receiving Device Through the Natural Media, of which the following is a specification, reference being had to the accompanying drawings, which form a part of the same.

The subject of my present invention is an improvement in the art of utilizing effects transmitted from a distance to a receiving device through the natural media; and it consists in a novel method hereinafter described.

My invention is particularly useful in connection with methods and apparatus for operating distant receiving devices by means of electrical disturbances produced by proper transmitters and conveyed to such receiving devices through the natural media; but it obviously has a wider range of applicability and may be employed, for example, in the investigation or utilization of terrestrial, solar, or other disturbances produced by natural causes.

Several ways or methods of transmitting electrical disturbances through the natural media and utilizing them to operate distant receivers are now known and have been applied with more or less success for accomplishing a variety of useful results. One of these ways consists in producing by a suitable apparatus rays or radiations—that is, disturbances—which are propagated in straight lines through space, directing them upon a receiving or recording apparatus at a distance, and thereby bringing the latter into action. This method is the oldest and best known and has been brought particularly into prominence in recent years through the investigations of Heinrich Hertz. Another method consists in passing a current through a circuit, preferably one inclosing a very large area, inducing thereby in a similar circuit situated at a distance another current and affecting by the same in any convenient way a receiving device. Still another way, which has also been known for many years, is to pass in any suitable manner a current through a portion of the ground, as by connecting to two points of the same, preferably at a considerable distance from each other, the two terminals of a generator and to energize by a part of the current diffused through the earth a distant circuit, which is similarly arranged and grounded at two points widely apart and which is made to act upon a sensitive receiver. These various methods have their limitations, one especially, which is common to all, being that the receiving circuit or instrument must be maintained in a definite position with respect to the transmitting apparatus, which often imposes great disadvantages upon the use of the apparatus.

In several applications filed by me and patents granted to me I have disclosed other methods of accomplishing results of this nature which may be briefly described as follows: In one system the potential of a point or region of the earth is varied by imparting to it intermittent or alternating electrifications through one of the terminals of a suitable source of electrical disturbances, which to heighten the effect has its other terminal connected to an insulated body, preferably of large surface and at an elevation. The electrifications communicated to the earth spread in all directions through the same, reaching a distant circuit, which generally has its terminals arranged and connected similarly to those of the transmitting source and operates upon a highly-sensitive receiver. Another method is based upon the fact that the atmospheric air, which behaves as an excellent insulator to currents generated by ordinary apparatus, becomes a conductor under the influence of currents or impulses of enormously high electromotive force which I have devised means for generating. By such means air strata, which are easily accessible, are rendered available for the production of many desired effects at distances however great. This method, furthermore, allowed advantage to be taken of many of those improvements which are practicable in the ordinary systems of transmission involving the use of a metallic conductor.

Obviously whatever method be employed it is desirable that the disturbances produced by the transmitting apparatus should be as powerful as possible, and by the use of certain forms of high-frequency apparatus which

I have devised and which are now well known important practical advantages are in this respect secured. Furthermore, since in most cases the amount of energy conveyed to the distant circuit is but a minute fraction of the total energy emanating from the source it is necessary for the attainment of the best results that whatever the character of the receiver and the nature of the disturbances as much as possible of the energy conveyed should be made available for the operation of the receiver, and with this object in view I have heretofore, among other means, employed a receiving-circuit of high self-induction and very small resistance and of a period such as to vibrate in synchronism with the disturbances, whereby a number of separate impulses from the source were made to coöperate, thus magnifying the effect exerted upon and insuring the action of the receiving device. By these means decided advantages have been secured in many instances; but very often the improvement is either not applicable at all, or if so the gain is very slight. Evidently when the source is one producing a continuous pressure or delivering impulses of long duration it is impracticable to magnify the effects in this manner, and when, on the other hand, it is one furnishing short impulses of extreme rapidity of succession the advantage obtained in this way is insignificant, owing to the radiation and the unavoidable frictional waste in the receiving-circuit. These losses reduce greatly both the intensity and the number of the coöperative impulses, and since the initial intensity of each of these is necessarily limited only an insignificant amount of energy is thus made available for a single operation of the receiver. As this amount is consequently dependent on the energy conveyed to the receiver by one single impulse, it is evidently necessary to employ either a very large and costly and therefore objectionable transmitter or else to resort to the equally objectionable use of a receiving device too delicate and too easily deranged. Furthermore, the energy obtained through the coöperation of the impulses is in the form of extremely-rapid vibrations and because of this unsuitable for the operation of ordinary receivers, the more so as this form of energy imposes narrow restrictions in regard to the mode and time of its application to such devices. To overcome these and other limitations and disadvantages that have heretofore existed in such systems of transmission of signals or intelligence and to render possible an investigation of impulses or disturbances propagated through the natural media from any kind of source and their practical utilization for any purpose to which they are applicable, I have devised a novel method, which I have described in a pending application filed June 24, 1899, Serial No. 721,790, and which, broadly stated, consists in effecting during any desired time interval a storage of energy derived from such impulses and utilizing the potential energy so obtained for operating a receiving device.

My present invention is intended for the same general purposes, and it comprises a modified method and apparatus by means of which similar results may be obtained.

The chief feature which distinguishes my present from my former invention just referred to is that the energy stored is not, as in the former instance, obtained from the energy of the disturbances or effects transmitted from a distance, but from an independent source.

Expressed generally, my present method consists in charging a storage device with energy from an independent source controlling the charging of said device by the action of the effects or disturbances transmitted through the natural media and coincidentally using the stored energy for operating a receiving device.

A great variety of disturbances produced either by suitably-constructed transmitters or by natural causes are at present known to be propagated through the natural media, and there are also a variety of means or devices enabling energy to be stored, and in view of this I wish to say that I consider the utilization of any such disturbances and the employment of any of these means as within the scope of my present invention so long as the use of the general method hereinbefore stated is involved.

The best way of carrying out my invention which I at present know is to store electrical energy obtained from a suitable electrical generator in a condenser and to control the storage or the application of this energy by means of a sensitive device acted upon by the effects or disturbances, and thereby cause the operation of the receiver.

In the practical application of this method I usually proceed as follows: At any point where I desire to investigate or to utilize for any purpose effects or disturbances propagated through the natural media from any kind of source I provide a suitable generator of electricity—as, for example, a battery and a condenser—which I connect to the poles of the generator in series with a sensitive device capable of being modified in its electrical resistance or other property by the action of the disturbances emitted from the source. To the terminals of the condenser I connect the receiver which is to be operated in series with another device of suitable construction which performs the function of periodically discharging the condenser through the receiver at and during such intervals of time as may be best suitable for the purpose contemplated. This latter device may merely consist of two stationary electrodes separated by a feeble dielectric layer of minute thickness, but sufficient to greatly reduce or practically interrupt the current in the circuit under normal conditions, or it may comprise terminals one or more of which are movable and actuated by any suitable force and are adapted to be brought into and out of contact with each other in any convenient manner. The sensitive device may be any of the many devices of this kind which are known to be affected by the disturbances, impulses, or effects propagated through the media, and it may be of such a character that normally—that is, when not acted upon—it entirely prevents the passage of electricity from the generator to the condenser, or it may be such that it allows a gradual leaking through of the current and a charging of the condenser at a slow rate. In any case it will be seen that if the disturbances, of whatever nature they may be, cause an appreciable diminution in the electrical resistance of the sensitive device the current from the battery will pass more readily into the condenser, which will be charged at a more rapid rate, and consequently each of its discharges through the receiver, periodically effected by the special device before referred to which performs this function, will be stronger than normally—that is, when the sensitive device is not acted upon by the disturbances. Evidently, then, if the receiver be so adjusted that it does not respond to the comparatively feeble normal discharges of the condenser, if they should occur, but only to those stronger ones which take place upon the diminution of the resistance of the sensitive device, it will be operated only when this device is acted upon by the disturbances, thus making it possible to investigate and to utilize the latter for any desired purpose.

The general principle underlying my invention and the operation of the various devices used will be clearly understood by reference to the accompanying drawings, in which—

Figure 1 is a diagram illustrating a typical arrangement of apparatus which may be used in carrying my method into practice, and Figs. 2, 3, 4, and 5 similar diagrams of modified arrangements of apparatus for the same purpose.

In Fig. 1, C is a condenser, to the terminals T and T′ of which is connected a charging-circuit including a battery B, a sensitive device *a*, and a resistance *r*, all connected in series, as illustrated. The battery should be preferably of very constant electromotive force and of an intensity carefully determined to secure the best results. The resistance *r*, which may be a frictional or an inductive one, is not absolutely necessary; but it is of advantage to use it in order to facilitate adjustment, and for this purpose it may be made variable in any convenient and preferably continuous manner. Assuming that the disturbances which are to be investigated or utilized for some practical end are rays identical with or resembling those of ordinary light, the sensitive device *a* may be a selenium cell properly prepared, so as to be highly susceptible to the influence of the rays, the action of which should be intensified by the use of a reflector A, shown in the drawings. It is well known that when cells of this kind are exposed to such rays of greatly-varying intensity they undergo corresponding modifications of their electrical resistance; but in the ways they have been heretofore used they have been of very limited utility. In addition to the circuit including the sensitive device or cell *a* another circuit is provided, which is likewise connected to the terminals T T′ of the condenser. This circuit, which may be called the "receiving-circuit," includes the receiver R and in series with it a device *d*, before referred to, which performs the duty of periodically discharging the condenser through the receiver. It will be noted that, as shown in Fig. 1, the receiving-circuit is in permanent connection with the battery and condenser terminal T, and it should be stated that it is sometimes desirable to entirely insulate the receiving-circuit at all times except the moments when the device *d* operates to discharge the condenser, thus preventing any disturbing influence which might otherwise be caused in this circuit by the battery or the condenser during the period when the receiver should not be acted upon. In such a case two devices, as *d*, may be used—one in each connection from the condenser to the receiving-circuit—or else one single device of this kind, but of a suitably-modified construction, so that it will make and break simultaneously and at proper intervals of time both of the connections of this circuit with the condenser T and T′.

From the foregoing the operation of the apparatus as illustrated in Fig. 1 will be at once understood. Normally—that is, when it is not influenced by the rays at all or very slightly—the cell *a* being of a comparatively high resistance permits only a relatively feeble current to pass from the battery into the condenser, and hence the latter is charged at too slow a rate to accumulate during the time interval between two succeeding operations of the device *d* sufficient energy to operate the receiver or, generally speaking, to produce the required change in the receiving-circuit. This condition is readily secured by a proper selection and adjustment of the various devices described, so that the receiver will remain unresponsive to the feeble discharges of the condenser which may take place when the cell *a* is acted upon but slightly or not at all by the rays or disturbances; but if now new rays are permitted to fall upon the cell or if the intensity of those already acting upon it be increased by any cause then its resistance will be diminished and the condenser will be charged by the battery at a more rapid rate, enabling sufficient potential energy to be stored in the condenser during the period of inaction of the device *d* to operate the receiver or to bring about any desired change in the receiving-circuit when the device *d* acts. If the rays acting upon the cell or sensitive device *a* are varied or intermitted in any arbitrary manner, as whe transmitting intelligence in the usual way from a distant station by means of short and long signals, the apparatus may readily be made to record or to enable an operator to read the message, since the receiver, supposing it to be an ordinary magnetic relay, for example, will be operated by each signal from the sending-station a certain number of times having some relation to the duration of each signal. It will be readily seen, however, that if the rays are varied in any other way, as by impressing upon them changes in intensity, the succeeding condenser discharges will undergo corresponding changes in intensity, which may be indicated or recorded by a suitable receiver and distinguished irrespectively of duration.

With reference to Fig. 1, it may be useful to state that the electrical connections of the various devices illustrated may be made in many different ways. For instance, the sensitive device instead of being in series, as shown, may be in a shunt to the condenser, this modification being illustrated in Fig. 3, in which the devices already described are indicated by similar letters to correspond with those of Fig. 1. In this case it will be observed that the condenser which is being charged from the battery B through the resistance *r*, preferably inductive and properly related to the capacity of the condenser, will store less energy when the sensitive device *a* is energized by the rays and its resistance thereby diminished. The adjustment of the various instruments may then be such that the receiver will be operated only when the rays are diminished in intensity or interrupted and entirely prevented from falling upon the sensitive cell, or the sensitive device may be placed, as shown in Fig. 4, in a shunt to the resistance *r* or inserted in any suitable way in the circuit containing the receiver—for example, as illustrated in Fig. 5—in both of which figures the various devices are lettered to correspond with those in Fig. 1, so that the figures become self-explanatory. Again, the several instruments may be connected in the manner of a Wheatstone bridge, as will be hereinafter explained with reference to Fig. 2, or otherwise connected or related; but in each case the sensitive device will have the same duty to perform—that is, to control the energy stored and utilized in some suitable way for causing the operation of the receiver in correspondence with the intermittences or variations of the effects or disturbances, and in each instance by a judicious selection of the devices and careful adjustment the advantages of my method may be more or less completely secured. I find it preferable, however, to follow the plan which I have illustrated and described.

It will be observed that the condenser is an important element in the combination. I have shown that by reason of its unique properties it greatly adds to the efficacy of this method. It allows the energy accumulated in it to be discharged instantaneously, and therefore in a highly-effective manner. It magnifies in a large degree the current supplied from the battery, and owing to these features it permits energy to be stored and discharged at practically any rate desired, and thereby makes it possible to obtain in the receiving-circuit very great changes of the current strength by impressing upon the battery-current very small variations. Other means of storage possessing these characteristics to a useful degree may be employed without departing from the broad spirit of my invention; but I prefer to use a condenser, since in these respects it excels any other storage device of which I have knowledge.

In Fig. 2 a modified arrangement of apparatus is illustrated which is particularly adapted for the investigation and utilization of very feeble impulses or disturbances, such as may be used in conveying signals or producing other desired effects at very great distances. In this case the energy stored in the condenser is passed through the primary of a transformer the secondary circuit of which contains the receiver, and in order to render the apparatus still more suitable for use in detecting feeble impulses in addition to the sensitive device which is acted upon by the impulses another such device is included in the secondary circuit of the transformer. The scheme of connections is in the main that of a Wheatstone bridge the four branches of which are formed by the sensitive device *a* and resistances L, L′, and L″, all of which should be preferably inductive and also adjustable in a continuous manner or at least by very small steps. The condenser C′, which is generally made of considerable capacity, is connected to two opposite points of the bridge, while a battery B, in series with a continuously-adjustable non-inductive resistance *r*′, is connected to the other pair of opposite points, as usual. The four resistances included in the branches of the bridge—namely, *a*, L, L′, and L″—are of a suitable size and so proportioned that under normal conditions—that is, when the device *a* is not influenced at all or only slightly by the disturbances—there will be no difference of potential or in any case the minimum of the same at the terminals T and T′ of the condenser. It is assumed in the present instance that the disturbances to be investigated or utilized are such as will produce a difference of electric potential, however small, between two points or regions in the natural media—as the earth, the water, or the air—and in order to apply this potential difference effectively to the sensitive device *a* the terminals of the same are connected to two plates P and P′, which should be of as large a surface as practicable and so located in the media that the largest possible difference of potential will be produced by the disturbances between the terminals of the sensitive device. This device is in the present case one of familiar construction, consisting of an insulating-tube, which is indicated by the heavy lines in the drawings and which has its ends closed tightly by two conducting-plugs with reduced extensions, upon which bear two brushes b b, through which the currents are conveyed to the device. The tubular space between the plugs is partially filled with a conducting sensitive powder, as indicated, the proper amount of the same and the size of its grains being determined and adjusted beforehand by experiment. This tube I rotate by clockwork or other means at a uniform and suitable rate of speed, and under these conditions I find that this device behaves toward disturbances of the kind before assumed in a manner similar to that of a stationary cell of selenium toward rays of light. Its electrical resistance is diminished when it is acted upon by the disturbances and is automatically restored upon the cessation of their influence. It is of advantage to employ round grains of powder in the tube, and in any event it is important that they should be of as uniform size and shape as possible and that provision should be made for maintaining an unchanging and very dry atmosphere in the tube. To the terminals T and T′ of the condenser C′ is connected a coil p, usually consisting of a few turns of a conductor of very small resistance, which is the primary of the transformer before referred to, in series with a device d, which effects the discharge of the condenser through the coil p at predetermined intervals of time. In the present case this device consists of a cylinder made partly of conducting and partly of insulating material e and e', respectively, which is rotated at the desired rate of speed by any suitable means. The conducting part e is in good electrical connection with shaft S and is provided with tapering segments, as f, upon which slides a brush k, which should preferably be capable of longitudinal adjustment along the cylinder. Another brush b', which is connected to the condenser-terminal T′, being arranged to bear upon the shaft S, it will be seen that whenever the brush k comes in contact with a conducting-segment f the circuit including the primary p will be completed and the condenser, if energized, discharged through the same. By an adjustment of the speed of rotation of the cylinder and a displacement of the brush k along the axis of the same the circuit may be made to open and close in as rapid succession and remain open or closed during such intervals of time as may be desired. In inductive relation to the primary p is a secondary coil s, usually of much thinner wire and of many more turns than the former, to which are connected in a series a receiver R, (illustrated as an ordinary magnetic relay,) a continuously-adjustable non-inductive resistance r'', a battery B′ of a properly determined and very constant electromotive force, and finally a sensitive device a' of the same or similar construction as a, which is likewise rotated at a uniform speed and which with its brushes b'' b'' closes the secondary circuit. The electromotive force of the battery B′ is so graduated by means of the adjustable resistance r'' that the dielectric layers in the sensitive device a' are strained very nearly to the point of breaking down and give way upon a slight increase of the electrical pressure on the terminals of the device. It will of course be understood that the resistance r'' is used mainly because of convenience and that it may be dispensed with, in which case the adjustment may be effected in many other ways, as by determining the proper amount or coarseness of the sensitive powder or by varying the distance apart of the metallic plugs in the ends of the tube. The same may be said of the resistance r', which is in series with the battery B and serves to graduate the force of the latter, so that the dielectric layers of the sensitive device a are subjected to a similar strain and maintained in a state of delicate poise. The various instruments being connected and adjusted in the manner described, it will now be readily seen from the foregoing that under normal conditions, the device a being unaffected by the disturbances, or practically so, and there being no or only a very insignificant amount of energy stored in the condenser, the periodical closure of the primary circuit of the transformer through the operation of the device d will have no appreciable effect upon the primary coil p, and hence no currents will be generated in the secondary coil s, at least not such as would disturb the state of delicate balance existing in the secondary circuit including the receiver, and therefore the latter will not be actuated by the battery B′; but when, owing to the disturbances or impulses propagated through the media from a distant source, an additional electromotive force, however small, is created between the terminals of the device a the dielectric layers in the same, unable to support the increased strain, give way and allow the current of the battery B to pass through, thus causing a difference of potential at the terminals T and T′ of the condenser. A sufficient amount of energy being now stored in this instrument during the time interval between each two succeeding operations of the device d, each closure of the primary circuit by the latter results in the passage of a sudden current impulse through the coil p, which induces a corresponding current of relatively high electromotive force in the secondary coil s. Owing to this the dielectric in the device a' gives way, and the current of the battery B′ being allowed to pass the receiver R is operated, but only for a moment, since by the rotation of the devices a, a', and d, which may be all driven from the same shaft, the original conditions are restored, assuming, of course, that the electromotive force set up by the disturbances at the terminals of the sensitive device *a* is only momentary or of a duration not longer than the time of closure of the primary circuit; otherwise the receiver will be actuated a number of times and so long as the influence of the disturbances upon the device *a* continues. In order to render the discharged energy of the condenser more effective in causing the operation of the receiver, the resistance of the primary circuit should be very small and the secondary coil *s* should have a number of turns many times greater than that of the primary coil *p*. It will be noted that since the condenser under the above assumptions is always charged in the same direction the strongest current impulse in the secondary coil, which is induced at the moment when the brush *k* comes in contact with segment *f*, is also of unchanging direction, and for the attainment of the best results it is necessary to connect the secondary coil so that the electromotive force of this impulse will be added to that of the battery and will momentarily strengthen the same. However, under certain conditions, which are well understood by those skilled in the art, the devices will operate whichever way the secondary be connected. It is preferable to make the inductive resistances L and L' relatively large, as they are in a shunt to the device *a* and might if made too small impair its sensitiveness. On the other hand, the resistance L'' should not be too large and should be related to the capacity of the condenser and the number of makes and breaks effected by the device *d* in well-known ways. Similar considerations apply, of course, to the circuits including the primary *p* and secondary *s*, respectively.

By carefully observing well-known rules of scientific design and adjustment of the instruments the apparatus may be made extremely sensitive and capable of responding to the feeblest influences, thus making it possible to utilize impulses or disturbances transmitted from very great distances and too feeble to be detected or utilized in any of the ways heretofore known, and on this account the method here described lends itself to many scientific and practical uses of great value. Obviously the character of the devices and the manner in which they are connected or related may be greatly varied without departing from the spirit of my invention.

What I claim as new, and desire to secure by Letters Patent, is—

1. The method hereinbefore described of utilizing effects or disturbances transmitted through the natural media, which consists in charging a storage device with energy from an independent source, controlling the charging of said device by the action of the effects or disturbances, and coincidently using the stored energy for operating a receiving device.

2. The method hereinbefore described of utilizing effects or disturbances transmitted from a distant source, which consists in charging the storage device with electrical energy from an independent source, controlling the charging of said device by the action of the effects or disturbances, and coincidently using the stored electrical energy for operating the receiving device.

3. The method hereinbefore described of utilizing effects or disturbances transmitted through the natural media, which consists in controlling, by means of such effects or disturbances, the charging of an electrical storage device from an independent source and discharging the stored energy through a receiving-circuit.

4. The method hereinbefore described of utilizing effects or disturbances transmitted through the natural media, which consists in controlling, by means of such effects or disturbances, the charging of an electrical condenser from an independent source, and discharging the stored energy through a receiving-circuit.

5. The method hereinbefore described of utilizing effects or disturbances transmitted through the natural media, which consists in effecting a storage during any desired time interval and under control of such effects or disturbances, of energy derived from an independent source, and utilizing the potential energy so obtained for operating a receiving device.

6. The method hereinbefore described of utilizing effects or disturbances transmitted through the natural media, which consists in effecting a storage, during any desired time interval and under the control of such disturbances or effects of electrical energy derived from an independent source, and utilizing the potential energy so obtained for operating a receiving device.

7. The method hereinbefore described of utilizing effects or disturbances transmitted through the natural media, which consists in effecting a storage in a condenser during any desired time interval and under the control of such disturbances or effects, of electrical energy derived from an independent source, and utilizing the potential energy so obtained for operating a receiving device.

8. The method hereinbefore described of utilizing effects or disturbances transmitted through the natural media from a distant source, which consists in storing, during succeeding intervals of time determined by means of such effects or disturbances, electrical energy derived from an independent source, and utilizing the potential energy so accumulated to operate a receiving device.

9. The method hereinbefore described of utilizing effects or disturbances transmitted through the natural media from a distant source, which consists in storing in a condenser during succeeding intervals of time determined by means of such effects or disturbances, electrical energy derived from an independent source, and utilizing the poten-

tial energy so accumulated to operate a receiving device.

10. The method hereinbefore described of utilizing effects or disturbances transmitted through the natural media from a distant source, which consists in storing, during succeeding intervals of time determined by means of such effects or disturbances, electrical energy derived from an independent source, and using, for periods of time predetermined as to succession and duration, the accumulated energy so obtained to operate a receiving device.

11. The method hereinbefore described of utilizing effects or disturbances transmitted through the natural media from a distant source, which consists in storing in a condenser during succeeding intervals of time determined by means of such effects or disturbances, electrical energy derived from an independent source, and using, for periods of time predetermined as to succession and duration, the accumulated energy so obtained to operate a receiving device.

12. The method hereinbefore described of utilizing electrical effects or disturbances transmitted through the natural media from a distant source, which consists in effecting by means of such disturbances or effects a storage in a storage device of electrical energy derived from an independent source for periods of time corresponding in succession and duration to such disturbances or effects, and discharging the electrical energy so accumulated into or through a receiving device at predetermined intervals of time.

13. The method hereinbefore described of utilizing electrical effects or disturbances transmitted from a distant source, which consists in effecting by means of such disturbances or effects a storage in a condenser of electrical energy derived from an independent source for periods of time corresponding in succession and duration to such disturbances or effects, and discharging the electrical energy so accumulated into or through a receiving device at predetermined intervals of time.

14. The method hereinbefore described of utilizing electrical effects or disturbances transmitted from a distant source, which consists in producing, by means of such effects or disturbances, variations of resistance in a circuit including an independent electrical source and a device adapted to be charged with electrical energy therefrom, thereby causing the storage device to be charged with energy from such independent source, and using the potential electrical energy so accumulated to operate a receiving device.

15. The method hereinbefore described of utilizing effects or disturbances transmitted through the natural media from a distant source, which consists in producing, by means of such effects or disturbances, variations of resistance in a circuit including an independent electrical source and a condenser, thereby causing the condenser to be charged with energy from the independent source, and using the potential electrical energy so accumulated to operate a receiving device.

16. The method hereinbefore described of utilizing effects or disturbances transmitted through the natural media from a distant source, which consists in causing, by means of such effects or disturbances, electrical energy from an independent source to be stored in a storage device, using the electrical energy so accumulated to operate a transformer and employing the secondary currents from such transformer to operate a receiving device.

17. The method hereinbefore described of utilizing effects or disturbances transmitted through the natural media from a distant source, which consists in causing, by means of such effects or disturbances, electrical energy from an independent source to be stored in a condenser, using the electrical energy so accumulated to operate a transformer and employing the secondary currents from such transformer to operate a receiving device.

18. The method hereinbefore described of utilizing effects or disturbances transmitted through the natural media from a distant source, which consists in causing, by means of such disturbances, variations of resistance in a circuit including an independent source of electricity and a storage device and thereby causing the storage device to be charged from such independent source, discharging the energy so accumulated in the storage device through the primary of a transformer at predetermined intervals of time, and operating a receiver by the currents so developed in the secondary of the transformer.

19. The method hereinbefore described of utilizing effects or disturbances transmitted through the natural media from a distant source, which consists in causing, by means of such disturbances, variations of resistance in a circuit including an independent source of electricity and a condenser and thereby causing the condenser to be charged from such independent source, discharging the energy so accumulated in the condenser through the primary of a transformer at predetermined intervals of time and operating a receiver by the currents so developed in the secondary of the transformer.

NIKOLA TESLA.

Witnesses:

F. LÖWENSTEIN,

E. A. SUNDERLIN.

No. 685,955. Patented Nov. 5, 1901.

N. TESLA.

APPARATUS FOR UTILIZING EFFECTS TRANSMITTED FROM A DISTANCE TO A RECEIVING DEVICE THROUGH NATURAL MEDIA.

(Application filed Sept. 8, 1899. Renewed May 29, 1901.)

(No Model.)

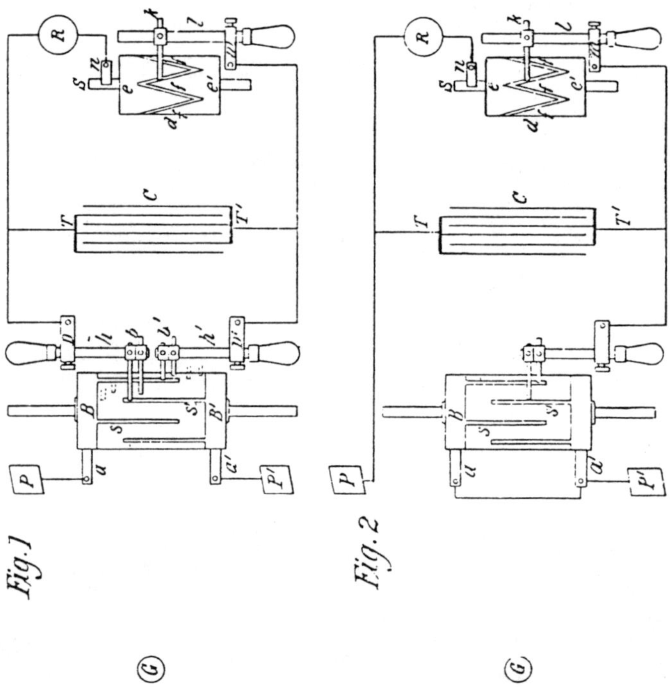

Witnesses:
G. B. Lsmo.
Hillary C. Messimer

Nikola Tesla, Inventor
by Ken. Page & Cooper.
Att'ys

UNITED STATES PATENT OFFICE.

NIKOLA TESLA, OF NEW YORK, N.Y.

APPARATUS FOR UTILIZING EFFECTS TRANSMITTED FROM A DISTANCE TO A RECEIVING DEVICE THROUGH NATURAL MEDIA.

SPECIFICATION forming part of Letters Patent No. 685,955, dated November 5, 1901.

Application filed June 24, 1899. Serial No. 721,790. Divided and this application filed September 8, 1899. Renewed May 29, 1901. Serial No. 62,317. (No model.)

To all whom it may concern:

Be it known that I, NIKOLA TESLA, a citizen of the United States, residing at the borough of Manhattan, in the city, county, and State of New York, have invented certain new and useful Improvements in Apparatus for Utilizing Effects Transmitted from a Distance to a Receiving Device Through the Natural Media, of which the following is a specification, reference being had to the accompanying drawings, whic form a part of the same.

This application is a division of an application filed by me June 24, 1899, Serial No. 721,790, in which a method of utilizing effects or disturbances transmitted through the natural media from a distant source is described and made the subject of the claims. The invention of my present application consists in the apparatus hereinafter described and claimed, by the use of which the method claimed in my said prior application may be practiced and by means of which results hitherto unattainable may be secured.

Several ways or methods of transmitting electrical disturbances through the natural media and utilizing them to operate distant receivers are now known and have been applied with more or less success for accomplishing a variety of useful results. One of these ways consists in producing by a suitable apparatus rays or radiations — that is, disturbances — which are propagated in straight lines through space, directing them upon a receiving or recording apparatus at a distance, and thereby bringing the latter into action. This method is the oldest and best known, and has been brought particularly into prominence in recent years through the investigations of Heinrich Hertz. Another method consists in passing a current through a circuit, preferably one inclosing a very large area, inducing thereby in a similar circuit, situated at a distance, another current and affecting by the same in any convenient way a receiving device. Still another way, which has also been known for many years, is to pass in any suitable manner a current through a portion of the ground, as by connecting to two points of the same, preferably at a considerable distance from each other, the two terminals of a generator and to energize by a part of the current diffused through the earth a distant circuit, which is similarly arranged and grounded at two points widely apart and which is made to act upon a sensitive receiver. These various methods have their limitations, one, especially, which is common to all, being that the receiving circuit or instrument must be maintained in a definite position with respect to the transmitting apparatus, which often imposes great disadvantages upon the use of the apparatus.

In several applications filed by me and patents granted to me I have disclosed other methods of accomplishing results of this nature, which may be briefly described as follows: In one system the potential of a point or region of the earth is varied by imparting to it intermittent or alternating electrifications through one of the terminals of a suitable source of electrical disturbances, which to heighten the effect has its other terminal connected to an insulated body, preferably of large surface and at an elevation. The electrificatios communicated to the earth spread in all directions through the same, reaching a distant circuit, which generally has its terminals arranged and connected similarly to those of the transmitting source, and operates upon a highly-sensitive receiver. Another method is based upon the fact that the atmospheric air, which behaves as an excellent insulator to currents generated by ordinary apparatus, becomes a conductor under the influence of currents of impulses of enormously high electromotive force which I have devised means for generating. by such means air strata, which are easily accessible, are rendered available for the production of many desired effects at distances, however great. This method, furthermore, allows advantage to be taken of many of those improvements which are practicable in the ordinary systems of transmission involving the use of a metallic conductor.

Obviously whatever method be employed it is desirable that the disturbances produced by the transmitting apparatus should be as powerful as possible, and by the use of certain forms of high-frequency apparatus which I have devised and which are now well known important practical advantages are in this respect secured. Furthermore, since in most cases the amount of energy conveyed to the distant circuit is but a minute fraction of the total energy emanating from the source, it is necessary for the attainment of the best results that whatever the character of the receiver and the nature of the disturbances as much as possible of the energy conveyed should be made available for the operation of the receiver, and with this object in view I have heretofore, among other means, employed a receiving-circuit of high self-induction and very small resistance and of a period such as to vibrate in synchronism with the disturbances, whereby a number of separate impulses from the source were made to cooperate, thus magnifying the effect exerted upon and insuring the action of the receiving device. By these means decided advantages have been secured in many instances; but very often the improvement is either not applicable at all or if so the gain is very slight. Evidently when the source is one producing a continuous pressure or delivering impulses of long duration it is impracticable to magnify the effects in this manner, and when, on the other hand, it is one furnishing short impulses of extreme rapidity of succession the advantage obtained in this way is insignificant, owing to the radiation and the unavoidable frictional waste in the receiving-circuit. These losses reduce greatly both the intensity and the number of the cooperative impulses, and since the initial intensity of each of these is necessarily limited only an insignificant amount of energy is thus made available for a single operation of the receiver. As this amount is consequently dependent on the energy conveyed to the receiver by one single impulse, it is evidently necessary to employ either a very large and costly, and therefore objectionable transmitter, or else resort to the equally objectionable use of a receiving device too delicate and too easily deranged. Furthermore, the energy obtained through the cooperation of the impulses is in the form of extremely-rapid vibrations and because of this unsuitable for the operation of ordinary receivers, the more so as this form of energy imposes narrow restrictions in regard to the mode and time of its application to such devices. To overcome these and other limitations and disadvantages which have heretofore existed in such systems of transmission of signals or intelligence is the object of my invention, which comprises a novel form of apparatus for accomplishing these results.

The apparatus which is employed at the receiving-station, described in general terms, consists in the combination of a storage device included in a circuit connecting points at a distance from the source of the disturbances and between which a difference of potential is created by such disturbances, a receiving-circuit connected with the storage device, a receiver included in such receiving-circuit, and means for closing the receiving-circuit at any desired moment, and thereby causing the receiver to be operated by the energy with which the storage device has been charged.

The best form of apparatus for carrying out my invention of which I am now aware and the manner of using the same will be understood from the following description and the accompanying drawings, in which —

Figure 1 is a diagrammatic illustration of such apparatus, and Fig. 2 a modified form or arrangement of the same.

At any two points in the transmitting medium between which there exists or may be obtained in any manner through the action of the disturbances or effects to be investigated or utilized a difference of electrical potential of any magnitude I arrange two plates or electrodes so that they may be oppositely charged through the agency of such effects or disturbances, and I connect these electrodes to the terminals of a highly-insulated condenser, generally of considerable capacity. To the condenser-terminals I also connect the receiver to be operated in series with a device of suitable construction which performs the function of periodically discharging the condenser through the receiver at and during such intervals of time as may be best suitable for the purpose contemplated. This device may merely consist of two stationary electrodes separated by a feeble dielectric layer of minute thickness, or it may comprise terminals one or more of which are movable and actuated by any suitable force and are adapted to be brought into and out of contact with each other in any convenient manner. It will now be readily seen that if the disturbances, of whatever nature they may be, cause definite amounts of electricity of the same sign to be conveyed to each of the plates or electrodes above mentioned either continuously or at intervals of time which are sufficiently long the condenser will be charged to a certain potential and an adequate amount of energy being thus stored during the time determined by the device effecting the discharge of the condenser the receiver will be periodically operated by the electrical energy so accumulated; but very often the character of the impulses and the conditions of their use are such that without further provision not enough potential energy would be accumulated in the condenser to operate the receiving device. This is the case when, for example, each of the plates or terminals receives electricity of rapidly-changing sign or even when each receives electricity of the same sign, but only during periods of which are short as compared with the intervals separating them. In such instances I resort to the use of a special device which I insert in the circuit between the plates and the condenser for the purpose of conveying to each of the terminals of the latter electrical charges of the proper quality and order of succession to enable the required amount of potential energy to be stored in the condenser.

There are a number of well-known devices, either without any moving parts or terminals or with elements reciprocated or rotated by the application of a suitable force, which offer a more ready passage to impulses of one sign or direction than to those of the other or permit only impulses of one kind or order of succession to traverse a path, and any of these or similar devices capable of fulfilling the requirements may be used in carrying my invention into practice. One such device of familiar construction which will serve to convey a clear understanding of this part of my invention and enable a person skilled in the art to apply the same is illustrated in the annexed drawings. It consists of a cylinder A, of insulating material, which is moved at a uniform rate of speed by clockwork or other suitable motive power and is provided with two metal rings B B′, upon which bear brushes *a* and *a*′, which are connected, respectively, in the manner shown to the terminal plates P and P′, above referred to. From the rings B B′ extend narrow metallic segments *s* and *s*′, which by the rotation of the cylinder A are brought alternately into contact with double brushes *b* and *b*′, carried by and in contact with conducting-holders *h* and *h*′, which are adjustable longitudinally in the metallic supports D and D′, as shown. The latter are connected to the terminals T and T′ of a condenser C, and it should be understood that they are capable of angular displacement as ordinary brush-supports. The object of using two brushes, as *b* and *b*′, in each of the holders *h* and *h*′ is to vary at will the duration of the electric contact of the plates P and P′ with the terminals T and T′, to which is connected a receiving-circuit, including a receiver R and a device *d* of the kind above referred to, which performs the duty of closing the receiving-circuit at predetermined intervals of time and discharging the stored energy through the receiver. In the present case this device consists of a cylinder made partly of conducting and partly of insulating material *e* and *e*′, respectively, which is rotated at the desired rate of speed by any suitable means. The conducting part *e* is in good electrical connection with the shaft S and is provided with tapering segments *f f*, upon which slides a brush *k*, supported on a conducting-rod *l*, capable of longitudinal adjustment in a metallic support *m*. Another brush *n* is arranged to bear upon the shaft S, and it will be seen that whenever one of the segments *f* comes in contact with the brush *k* the circuit including the receiver R is completed and the condenser discharged through the same. By an adjustment of the speed of rotation of the cylinder *d* and a displacement of the brush *k* along the cylinder the circuit may be made to open and close in as rapid succession and remain open or closed during such intervals of time as may be desired. The plates P and P′ through which the electrifications are conveyed to the brushes *a* and *a*′ may be at a considerable distance from each other and both in the ground or both in the air or one in the ground and the other in the air, preferably at some height, or they may be connected to conductors extending to some distance or to the terminals of any kind of apparatus supplying electrical energy which is obtained from the energy of the impulses or disturbances transmitted from a distance through the natural media.

In illustration of the operation of the devices described let it be assumed that alternating electrical impulses from a distant generator, as G, are transmitted through the earth and that it is desired to utilize these impulses in accordance with my method. This may be the case, for example, when such a generator is used for purposes of signaling in one of the ways before enumerated, as by having its terminals connected at two points of the earth distant from each other. In this case the plates P and P′ are first connected to two properly-selected points of the earth, the speed of rotation of the cylinder A is varied until it is made to turn in synchronism with the alternate impulses of the generator, and, finally, the position of the brushes *b* and *b*′ is adjusted by angular displacement, as usual, or in other ways, so that they are in contact with the segments *s* and *s*′ during the periods when the impulses are at or near the maximum of their intensity. Only ordinary electrical skill and knowledge are required to make these adjustments, and a number of devices for effecting synchronous movement being well known and it being the chief object of my present application to set forth a novel apparatus embodying a general principle a detailed description of such devices is not considered necessary. I may state, however, that for practical purposes in the present case it is only necessary to shift the brushes back and forth until the maximum effect is secured. The above requirements being fulfilled, electrical charges of the same sign will be conveyed to each of the condenser-terminals as the cylinder A is rotated, and with each fresh impulse the condenser will be charged to a higher potential. The speed of rotation of the cylinder *d* being adjustable at will, the energy of any number of separate impulses may thus be accumulated in potential form and discharged through the receiver R upon the brush *k* coming in contact with one of the segments *f*. It will be of course understood that the capacity of the condenser should be such as to allow the storing of a much greater amount of energy than is required for the ordinary operation of the receiver. Since by this method a relatively great amount of energy and in a suitable form may be made available for the operation of a receiver, the latter need not be very sensitive but of course then the impulses are very feeble, as when coming from a great distance or when it is desired to operate a receiver very rapidly, than any of the well-known devices capable of responding to very feeble influences may be used in this connection.

If instead of the alternating impulses short impulses of the same direction are conveyed to the plates P and P′, the apparatus described may still readily be used, and for this purpose it is merely necessary to shift the brush *b* and *b′* into the position indicated by the dotted lines, while maintaining the same conditions in regard to synchronism as before, so that the succeeding impulses will be permitted to pass into the condenser, but prevented from returning to the ground or transmitting medium during the intervals between them, owing to the interruption during such intervals of the connections leading from the condenser-terminals to the plates.

Another way of using the apparatus with impulses of the same direction is the take off one pair of brushes, as *b*, disconnect the plate P from brush *a* and join it directly to the terminal T of the condenser, and to connect brush *a* with brush *a′*. When thus modified, the apparatus appears as shown in Fig. 2. Operated in this manner and assuming the speed of rotation of cylinder A to be the same, the apparatus will now be evidently adapted for a number of impulses per unit of time twice as great as in the preceding case. In all cases it is evidently important to adjust the duration of contact of segments *s* and *s′* with brushes *b b′* in the manner indicated.

When the apparatus I have described is used in connection with the transmission of signals or intelligence, it will of course be understood that the transmitter is operated is such a way as to produce disturbances or effects which are varied or intermitted in some arbitrary manner — for example, to produce longer and shorter successions of impulses, corresponding to the dashes and dots of the Morse alphabet — and the receiving device will respond to and indicate these variations or intermittences, since the storage device will be charged and discharged a number of times corresponding to the duration of the successions of impulses received.

Obviously the special appliances used in carrying out my invention may be varied in many ways without departing from the spirit of the same.

It is to be observed that it is the function of the cylinder A, with its brushes and connections to render the electrical impulses coming from the plates P and P′ suitable for charging the condenser (assuming them to be unsuitable for this purpose in the form in which they are received) by rectifying them when they are originally alternating in direction or by selecting such parts of them as are suitable when all are not, and any other device performing this function will obviously answer the purpose. It is also evident that a device such as I have already referred to which offers a more ready passage to impulses of one sign or permits only impulses of the same sign to pass may also be used to perform this selective function in many cases when alternating impulses are received. When the impulses are long and all of the same direction, and even when they are alternating but sufficiently long in duration and sustained in electromotive force, the brushes *b* and *b′* may be adjusted so as to bear on the parts B B′ of the cylinder A, or the cylinder and its brushes may be omitted and the terminals of the condenser connected directly to the plates P and P′

It will be seen that by the use of my invention results hitherto unattainable in utilizing disturbances or effects transmitted through natural media may be readily attained, since however great the distance of such transmission and however feeble or attenuated the impulses received enough energy may be accumulated from them by storing up the energy of succeeding impulses for a sufficient interval of time to render the sudden liberation of it highly effective in operating a receiver. In this way receivers of a variety of forms may be made to respond effectively to impulses too feeble to be detected or to be made to produce any sensible effect in any other way of which I am aware—a result of great value in scientific research as well as in various applications to practical use.

What I claim as my invention, and desire to secure by Letters Patent, is—

1. In an apparatus for utilizing electrical effects or disturbances transmitted through the natural media, the combination with a source of such effects or disturbances of a charging-circuit adapted to be energized by the action of such effects or disturbances, a storage device included in the charging-circuit and adapted to be charged thereby, a receiver, and means for causing the receiver to be operated by the energy accumulated in the storage device at arbitrary intervals of time, substantially as describewd.

2. In an apparatus for utilizing electrical effects or disturbances transmitted through the natural media, the combination with a source of such effects or disturbances of a charging-circuit adapted to be energized by the action of such effects or disturbances, a storage device included in the charging-circuit and adapted to be charged thereby, means for commutating, directing or selecting the current impulses in the charging-circuit, a receiving-circuit, and means for discharging the storage device through the receiving-circuit, substantially as described.

3. In an apparatus for utilizing electrical effects or disturbances transmitted through the natural media, the combination with a source of such effects or distrubances of a charging-circuit adapted to be energized by the action of such effects or disturbances, a condenser included in the charging-circuit and adapted to be charged thereby, means for commutating, directing or selecting the current impulses

in the charging-circuit, a receiving-circuit, and means for discharging the condenser through the receiving-circuit, substantially as described.

4. In an apparatus for utilizing electrical effects or disturbances transmitted through the natural media, the combination with a source of such effects or disturbances of a charging-circuit adapted to be energized by the action of such effects or disturbances, a storage device included in the charging-circuit and adapted to be charged thereby, means for commutating, directing or selecting the current impulses in the charging-circuit so as to render them suitable for charging the storage device, a receiving-circuit, and means for discharging the storage device thorugh the receiving-circuit, substantially as described.

5. In an apparatus for utilizing electrical effects or disturbances transmitted through the natural media, the combination with a source of such effects or distrubances of an charging-circuit adapted to be energized by the action of such effects or disturbances, a condenser included in the charging-circuit and adapted to be charged thereby, means for commutating, directing or selecting the current impulses in the charging-circuit so as to render them suitable for charging the condenser, a receiving-circuit, and means for dicharging the condenser thorugh the receiving-circuit, substantially as described.

6. In an apparatus for utilizing electrical effects or disturbances transmitted through the natural media, the combination with a source of such effects or disturbances of a charging-circuit adapted to be energized by the action of such effects or disturbances, a storage device included in the charging-circuit and adapted to be charged thereby, means for commutating, directing or selecting the current impulses in the charging-circuit so as to render them suitable for charging the storage device, a receiving-circuit, and means for discharging the storage device through the receiving-circuit at arbitrary intervals of time, substantially as described.

7. In an apparatus for utilizing electrical effects or disturbances transmitted to a distant receiving-station, the combination with a source of such effects or disturbances of a circuit distant from the source and adapted to have current impulses set up in it by the action of the effects or disturbances, a storage device, means for commutating, directing or selecting the impulses and connecting the circuit with the storage device at succeeding intervals of time synchronizing with the impulses, a receiving-circuit, and means for periodically discharging the storage device through the receiving-circuit, substantially as described.

8. In an apparatus for utilizing electrical effects or disturbances transmitted to a distant receiving-station, the combination with a source of such effects or disturbances of a circuit distant from the source and adapted to have current impulses set up in it by the action of the effects or disturbances, a condenser, means for commutating, directing or selecting the impulses and connecting the circuit with the condenser at succeeding intervals of time synchronizing with the impulses, a receiving-circuit, and means for periodically discharging the condenser through the receiving-circuit, substantially as described.

9. In an apparatus for utilizing electrical effects or disturbances transmitted through the natural media, the combination with a source of such effects or disturbances of a circuit connecting points at a distance from the source between which a difference of potential is created by such effects or disturbances, a storage device included in such circuit and adapted to be charged with the energy supplied by the same, a receiving-circuit connected with the storage device, a receiver included in such receiving-circuit, and means for closing the receiving-circuit and thereby causing the receiver to be operated by the energy accumulated in the storage device, substantially as described.

10. In an apparatus for utilizing electrical effects or disturbances transmitted through the natural media, the combination with a source of such effects or disturbances of a circuit at a distance from the source which is energized by such effects or disturbances, a storage device adapted to be charged with the energy supplied by such circuit, means for connecting the storage device with the said circuit for periods of time predetermined as to succession and duration, a receiving-circuit connected with the storage device, a receiver included in such receiving-circuit, and means for closing the receiving-circuit and thereby causing the receiver to be operated by the energy accumulated in the storage device, substantially as described.

11. In an apparatus for utilizing electrical effects or disturbances transmitted through the natural media, the combination of a circuit connecting points at a distance from the source between which a difference of potential is created by such effects or disturbances, a storage device included in such circuit and adapted to be charged with the energy supplied by the same, a receiving-circuit, a receiver included in such circuit, and means for connecting the receiving-circuit with the storage device for periods of time predetermined as to succession and duration and thereby causing the receiver to be operated by the energy accumulated in the storage device, substantially as described.

12. In an apparatus for utilizing electrical effects or disturbances transmitted thorugh the natural media, the combination of a circuit connecting points at a distance from the source between which a difference of potential is created by such effects or disturbances, a storage device adapted to be charged with the energy supplied by such circuit for succeeding and predetermined periods of time, a receiving-circuit, a receiver included in the

receiving-circuit, and means for connecting the receiving-circuit with the storage device for periods of time predetermined as to succession and duration and thereby causing the receiver to be operated by the energy accumulated in the storage device, substantially as described.

13. In an apparatus for utilizing electrical effects or disturbances transmitted through the natural media, the combination of a circuit connecting points at a distance from the source, between which a difference of potential is created by such effects or disturbances, a condenser included in such circuit and adapted to be charged by the current in the same, a receiving-circuit connected with the condenser, a receiver included in such receiving-circuit, and a device adapted to close the receiving-circuit at arbitrary intervals of time and thereby cause the receiver to be operated by the electrical energy accumulated in the condenser, substantially as described.

14. In an apparatus for utilizing electrical effects or disturbances transmitted through the natural media, the combination of a charging-circuit distant from the source and energized by the effects or disturbances, a storage device included in the charging-circuit, means included in the charging-circuit and acting in synchronism with the impulses therein for commutating, directing or selecting the impulses, a receiving-circuit and means for periodically discharging the storage device thorugh the receiving-circuit, substantially as described.

15. In an apparatus for utilizing electrical effects or disturbances transmitted through the natural media, the combination of a charging-circuit distant from the source and energized by the effects or disturbances, a condenser included in the charging-circuit, means included in the charging-circuit and acting in synchronism with the impulses therein for commutating, directing or selecting the impulses, a receiving-circuit and means for periodically discharging the condenser through the receiving-circuit, substantially as described.

16. In an apparatus for transmitting signals or intelligence through the natural media from a sending-station to a distant point, the combination of a generator or transmitter adapted to produce arbitrarily varied or intermitted electrical disturbances or effects in the natural media, a charging-circuit at the distant point adapted to receive corresponding electrical impulses or effects from the disturbances or effects so produced, a storage device included in the charging-circuit means included in the charging-circuit and acting in synchronism with the impulses therein for commutating, directing or selecting the impulses so as to render them suitable for charging the storage device, a receiving-circuit, substantially as described.

17. In an apparatus for transmitting signals or intelligence through the natural media from a sending-station to a distant point, the combination of a generator or transmitter adapted to produce arbitrarily varied or intermitted electrical disturbances or effects in the natural media, a charging-circuit at the distant point adapted to receive corresponding electrical impulses or effects from the disturbances or effects so produced, a condenser included in the charging-circuit, means included in the charging-circuit and acting in synchronism with the impulses therein for commutating, directing or selecting the impulses so as to render them suitable for charging the condenser, a receiving-circuit and means for periodically discharging the condenser through the receiving-circuit, substantially as described.

18. In an apparatus for transmitting signals or intelligence through the natural media from a sending-station to a distant point, the combination of a generator or transmitter adapted to produce arbitrarily varied or intermitted electrical disturbances or effects in the natural media, a circuit at the distant point adapted to receive corresponding electrical impulses or disturbances from the disturbances or effects so transmitted, a storage device included in such circuit and adapted to be charged thereby, a receiving-circuit connected with the storage device, a receiver included in the receiving-circuit and a device for closing the receiving-circuit at arbitrary intervals of time and thereby causing the receiver to be operated by the energy accumulated in the storage device, substantially as described.

NIKOLA TESLA.

Witnesses:

C.E. TITUS.

LEONARD E. CURTIS.

No. 685,956. Patented Nov. 5, 1901.

N. TESLA.

APPARATUS FOR UTILIZING EFFECTS TRANSMITTED THROUGH NATURAL MEDIA.

(Application filed Nov. 2, 1899. Renewed May 29, 1901.)

(No Model.)

Fig. 1

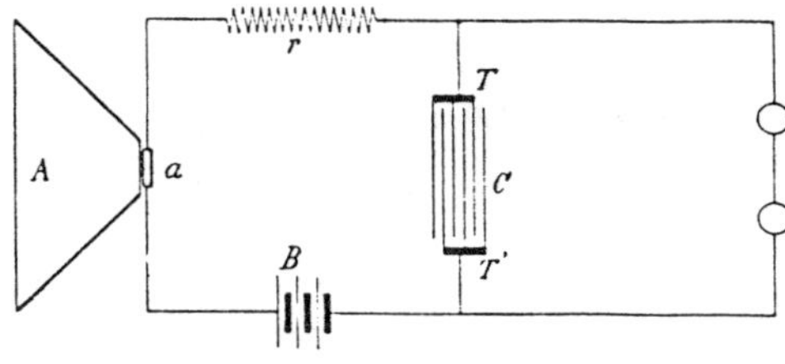

Fig. 2

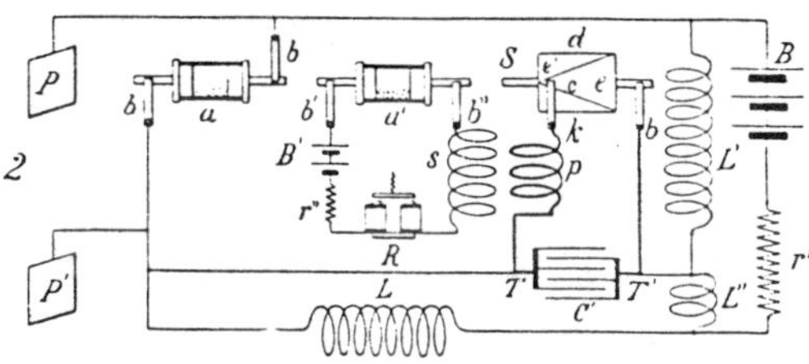

Fig. 3

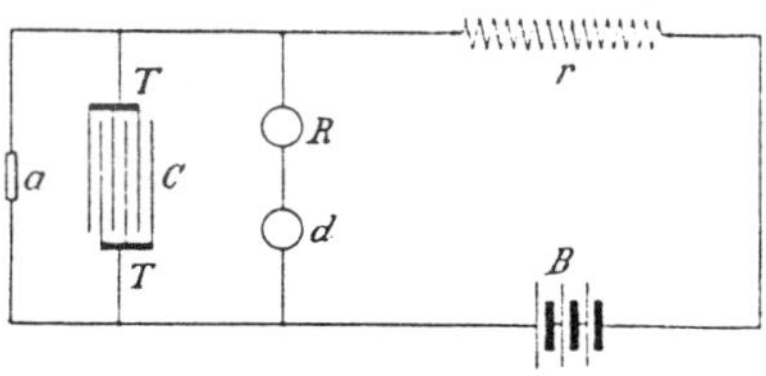

Fig. 4

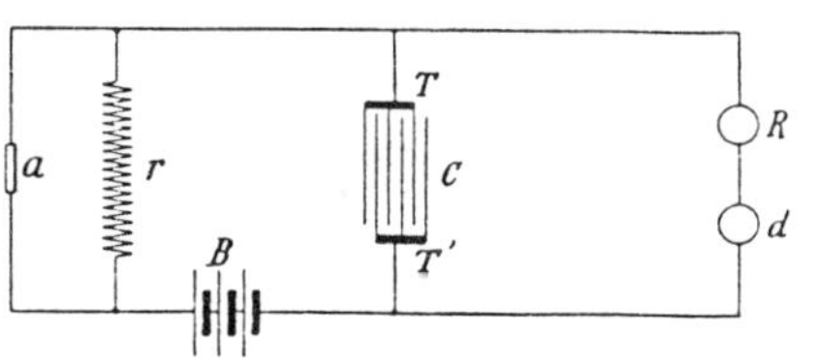

Fig. 5

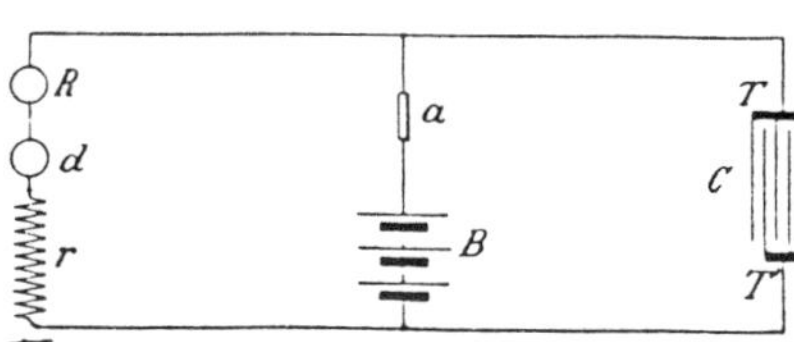

Witnesses:
Raphaël Netter
M. Lawson Dyer.

Inventor
Nikola Tesla

by Kerr, Page & Cooper, Attorneys.

UNITED STATES PATENT OFFICE.

NIKOLA TESLA, OF NEW YORK, N. Y.

APPARATUS FOR UTILIZING EFFECTS TRANSMITTED THROUGH NATURAL MEDIA.

SPECIFICATION forming part of Letters Patent No. 685,956, dated November 5, 1901.

Original application filed August 1, 1899, Serial No. 725,749. Divided and this application filed November 2, 1899. Renewed May 29, 1901. Serial No. 62,318. (No model.)

To all whom it may concern:

Be it known that I, NIKOLA TESLA, a citizen of the United States, residing at New York city, in the county and State of New York, have invented a new and useful Improvement in Apparatus for Utilizing Effects Transmitted from a Distance to a Receiving Device Through the Natural Media, of which the following is a specification, reference being had to the accompanying drawings, which form a part of the same.

The subject of my present invention is an improvement in the art of utilizing effects transmitted from a distance to a receiving device through the natural media; and it consists in the novel apparatus hereinafter described.

This application is a division of one filed by me August 1, 1899, Serial No. 725,749, and based upon and claiming the method herein described and which may be practiced by the use of the apparatus forming the subject of this application.

My invention is particularly useful in connection with methods and apparatus for operating distant receiving devices by means of electrical disturbances produced by proper transmitters and conveyed to such receiving devices through the natural media; but it obviously has a wider range of applicability and may be employed, for example, in the investigation or utilization of terrestrial, solar, or other disturbances produced by natural causes.

Several ways or methods of transmitting electrical disturbances through the natural media and utilizing them to operate distant receivers are now known and have been applied with more or less success for accomplishing a variety of useful results. One of these ways consists in producing by a suitable apparatus rays or radiations—that is, disturbances—which are propagated in straight lines through space, directing them upon a receiving or recording apparatus at a distance, and thereby bringing the latter into action. This method is the oldest and best known and has been brought particularly into prominence in recent years through the investigations of Heinrich Hertz. Another method consists in passing a current through a circuit, preferably one inclosing a very large area, inducing thereby in a similar circuit situated at a distance another current and affecting by the same in any convenient way a receiving device. Still another way, which has also been known for many years, is to pass in any suitable manner a current through a portion of the ground, as by connecting to two points of the same, preferably at a considerable distance from each other, the two terminals of a generator and to energize by a part of the current diffused through the earth a distant circuit, which is similarly arranged and grounded at two points widely apart and which is made to act upon a sensitive receiver. These various methods have their limitations, one especially, which is common to all, being that the receiving circuit or instrument must be maintained in a definite position with respect to the transmitting apparatus, which often imposes great disadvantages upon the use of the apparatus.

In several applications filed by me and patents granted to me I have disclosed other methods of accomplishing results of this nature, which may be briefly described as follows: In one system the potential of a point or region of the earth is varied by imparting to it intermittent or alternating electrifications through one of the terminals of a suitable source of electrical disturbances, which, to heighten the effect, has its other terminal connected to an insulated body, preferably of large surface and at an elevation. The electrifications communicated to the earth spread in all directions through the same, reaching a distant circuit, which generally has its terminals arranged and connected similarly to those of the transmitting source and operates upon a highly-sensitive receiver. Another method is based upon the fact that the atmospheric air, which behaves as an excellent insulator to currents generated by ordinary apparatus, becomes a conductor under the influence of currents or impulses of enormously high electromotive force, which I have devised means for generating. By such means air strata, which are easily accessible, are rendered available for the production of many desired effects at distances however great. This method, furthermore, allows advantage to be taken of many of those improvements which are practicable in the ordinary systems of transmission involving the use of a metallic conductor.

Obviously whatever method be employed it is desirable that the disturbances produced by the transmitting apparatus should be as powerful as possible, and by the use of certain forms of high-frequency apparatus, which I have devised and which are now well known, important practical advantages are in this respect secured. Furthermore, since in most cases the amount of energy conveyed to the distant circuit is but a minute fraction of the total energy emanating from the source it is necessary for the attainment of the best results that whatever the character of the receiver and the nature of the disturbances as much as possible of the energy conveyed should be made available for the operation of the receiver, and with this object in view I have heretofore, among other means, employed a receiving-circuit of high self-induction and very small resistance and of a period such as to vibrate in synchronism with the disturbances, whereby a number of separate impulses from the source were made to coöperate, thus magnifying the effect exerted upon and insuring the action of the receiving device. By these means decided advantages have been secured in many instances; but very often the improvement is either not applicable at all, or, if so, the gain is very slight. Evidently when the source is one producing a continuous pressure or delivering impulses of long duration it is impracticable to magnify the effects in this manner, and when, on the other hand, it is one furnishing short impulses of extreme rapidity of succession the advantage obtained in this way is insignificant, owing to the radiation and the unavoidable frictional waste in the receiving-circuit. These losses reduce greatly both the intensity and the number of the coöperative impulses, and since the initial intensity of each of these is necessarily limited only an insignificant amount of energy is thus made available for a single operation of the receiver. As this amount is consequently dependent on the energy conveyed to the receiver by one single impulse, it is evidently necessary to employ either a very large and costly, and therefore objectionable, transmitter, or else to resort to the equally objectionable use of a receiving device too delicate and too easily deranged. Furthermore, the energy obtained through the coöperation of the impulses is in the form of extremely rapid vibrations and because of this unsuitable for the operation of ordinary receivers, the more so as this form of energy imposes narrow restrictions in regard to the mode and time of its application to such devices. To overcome these and other limitations and disadvantages that have heretofore existed in such systems of transmission of signals or intelligence and to render possible an investigation of impulses or disturbances propagated through the natural media from any kind of source and their practical utilization for any purpose to which they are applicable, I have devised a novel method, which I have described in a pending application, filed June 24, 1899, Serial No. 721,790, and which, broadly stated, consists in effecting during any desired time interval a storage of energy derived from such impulses and utilizing the potential energy so obtained for operating a receiving device.

My present invention is intended for the same general purposes, and it comprises another apparatus by means of which similar results may be obtained.

The chief feature which distinguishes the method of my present from that of my former invention, just referred to, is that the energy stored is not, as in the former instance, obtained from the energy of the disturbances or effects transmitted from a distance, but from an independent source.

Expressed generally, the present method consists in charging a storage device with energy from an independent source, controlling the charging of said device by the action of the effects or disturbances transmitted through the natural media, and coincidently using the stored energy for operating a receiving device.

A great variety of disturbances, produced either by suitably-constructed transmitters or by natural causes, are at present known to be propagated through the natural media, and there are also a variety of means or devices enabling energy to be stored, and in view of this I wish to say that I consider the utilization of any such disturbances and the employment of any of these means as within the scope of my present invention so long as the use of the general method hereinbefore stated is involved.

The best way of carrying out my invention which I at present know is to store electrical energy obtained from a suitable electrical generator in a condenser and to control the storage or the application of this energy by means of a sensitive device acted upon by the effects or disturbances, and thereby cause the operation of the receiver.

In the practical application of this method I usually proceed as follows: At any point where I desire to investigate or to utilize for any purpose effects or disturbances propagated through the natural media from any kind of source I provide a suitable generator of electricity—as, for example, a battery and a condenser—which I connect to the poles of the generator in series with a sensitive device capable of being modified in its electrical resistance or other property by the action of the disturbances emitted from the source. To the terminals of the condenser I connect the receiver which is to be operated in series with another device of suitable construction, which performs the function of periodically discharging the condenser through the receiver at and during such intervals of time as may be best suitable for the purpose contemplated. This latter device may merely consist of two stationary electrodes separated by a feeble dielectric layer of minute thickness, but sufficient to greatly reduce or practically interrupt the current in the circuit under normal conditions, or it may comprise terminals one or more of which are movable and actuated by any suitable force and are adapted to be brought into and out of contact with each other in any convenient manner. The sensitive device may be any of the many devices of this kind which are known to be affected by the disturbances, impulses, or effects propagated through the media, and it may be of such a character that normally—that is, when not acted upon—it entirely prevents the passage of electricity from the generator to the condenser, or it may be such that it allows a gradual leaking through of the current and a charging of the condenser at a slow rate. In any case it will be seen that if the disturbances, of whatever nature they may be, cause an appreciable diminution in the electrical resistance of the sensitive device the current from the battery will pass more readily into the condenser, which will be charged at a more rapid rate, and consequently each of its discharges through the receiver, periodically effected by the special device before referred to which performs this function, will be stronger than normally—that is, when the sensitive device is not acted upon by the disturbances. Evidently then if the receiver be so adjusted that it does not respond to the comparatively feeble normal discharges of the condenser, if they should occur, but only to these stronger ones which take place upon the diminution of the resistance of the sensitive device it will be operated only when this device is acted upon by the disturbances, thus making it possible to investigate and to utilize the latter for any desired purpose.

The general principle underlying my invention and the operation of the various devices used will be clearly understood by reference to the accompanying drawings, in which—

Figure 1 is a diagram illustrating a typical arrangement of apparatus which may be used in carrying my method into practice, and Figs. 2, 3, 4, and 5 similar diagrams of modified arrangements of apparatus for the same purpose.

In Fig. 1, C is a condenser, to the terminals T and T′ of which is connected a charging-circuit, including a battery B, a sensitive device *a*, and a resistance *r*, all connected in series, as illustrated. The battery should be preferably of very constant electromotive force and of an intensity carefully determined to secure the best results. The resistance *r*, which may be a frictional or an inductive one, is not absolutely necessary; but it is of advantage to use it in order to facilitate adjustment, and for this purpose it may be made variable in any convenient and preferably continuous manner. Assuming that the disturbances which are to be investigated or utilized for some practical end are rays identical with or resembling those of ordinary light, the sensitive device *a* may be a selenium cell properly prepared, so as to be highly susceptible to the influence of the rays, the action of which should be intensified by the use of a reflector A. (Shown in the drawings.) It is well known that when cells of this kind are exposed to such rays of greatly-varying intensity they undergo corresponding modifications of their electrical resistance; but in the ways they have been heretofore used they have been of very limited utility.

In addition to the circuit including the sensitive device or cell *a* another circuit is provided, which is likewise connected to the terminals T T′ of the condenser. This circuit, which may be called the "receiving-circuit," includes the receiver R and in series with it a device *d*, before referred to, which performs the duty of periodically discharging the condenser through the receiver. It will be noted that, as shown in Fig. 1, the receiving-circuit is in permanent connection with the battery and condenser-terminal T, and it should be stated that it is sometimes desirable to entirely insulate the receiving-circuit at all times, except the moments when the device *d* operates to discharge the condenser, thus preventing any disturbing influence which might otherwise be caused in this circuit by the battery or the condenser during the period when the receiver should not be acted upon. In such a case two devices, as *d*, may be used, one in each connection from the condenser to the receiving-circuit, or else one single device of this kind, but of a suitably-modified construction, so that it will make and break simultaneously and at proper intervals of time both of the connections of this circuit with the condenser T and T′.

From the foregoing the operation of the apparatus, as illustrated in Fig. 1, will be at once understood. Normally—that is, when it is not influenced by the rays at all or very slightly—the cell *a*, being of a comparatively high resistance, permits only a relatively feeble current to pass from the battery into the condenser, and hence the latter is charged at too slow a rate to accumulate during the time interval between two succeeding operations of the device *d* sufficient energy to operate the receiver or, generally speaking, to produce the required change in the receiving-circuit. This condition is readily secured by a proper selection and adjustment of the various devices described, so that the receiver will remain unresponsive to the feeble discharges of the condenser which may take place when the cell *a* is acted upon but slightly or not at all by the rays or disturbances; but if now rays are permitted to fall upon the cell or if the intensity of those already acting upon it be increased by any cause then its resistance will be diminished and the condenser will be charged by the battery at a more rapid rate, enabling sufficient potential energy to be stored in the condenser during the period of inaction of the device *d* to operate the receiver or to bring about any desired change in the receiving-circuit when the device *d* acts. If the rays acting upon the cell or sensitive device *a* are varied or intermitted in any arbitrary manner, as when transmitting intelligence in the usual way from a distant station by means of short and long signals, the apparatus may readily be made to record or to enable an operator to read the message, since the receiver—supposing it to be an ordinary magnetic relay, for example—will be operated by each signal from the sending-station a certain number of times, having some relation to the duration of each signal. It will be readily seen, however, that if the rays are varied in any other way, as by impressing upon them changes in intensity, the succeeding condenser discharges will undergo corresponding changes in intensity, which may be indicated or recorded by a suitable receiver and distinguished irrespectively of duration.

With reference to Fig. 1 it may be useful to state that the electrical connections of the various devices illustrated may be made in many different ways. For instance, the sensitive device instead of being in series, as shown, may be in a shunt to the condenser, this modification being illustrated in Fig. 3, in which the devices already described are indicated by similar letters to correspond with those of Fig. 1. In this case it will be observed that the condenser, which is being charged from the battery B through the resistance *r*, preferably inductive and properly related to the capacity of the condenser, will store less energy when the sensitive device *a* is energized by the rays, and its resistance thereby diminished. The adjustment of the various instruments may then be such that the receiver will be operated only when the rays are diminished in intensity or interrupted and entirely prevented from falling upon the sensitive cell, or the sensitive device may be placed, as shown in Fig. 4, in a shunt to the resistance *r* or inserted in any suitable way in the circuit containing the receiver—for example, as illustrated in Fig. 5—in both of which figures the various devices are lettered to correspond with those in Fig. 1, so that the figures become self-explanatory. Again, the several instruments may be connected in the manner of a Wheatstone bridge, as will be hereinafter explained with reference to Fig. 2, or otherwise connected or related; but in each case the sensitive device will have the same duty to perform—that is, to control the energy stored and utilized in some suitable way for causing the operation of the receiver in correspondence with the intermittances or variations of the effects or disturbances—and in each instance by a judicious selection of the devices and careful adjustment the advantages of my method may be more or less completely secured. I find it preferable, however, to follow the plan which I have illustrated and described.

It will be observed that the condenser is an important element in the combination. I have shown that by reason of its unique properties it greatly adds to the efficacy of this method. It allows the energy accumulated in it to be discharged instantaneously, and therefore in a highly effective manner. It magnifies in a large degree the current supplied from the battery, and owing to these features it permits energy to be stored and discharged at practically any rate desired, and thereby makes it possible to obtain in the receiving-circuit very great changes of the current strength by impressing upon the battery-current very small variations. Other means of storage possessing these characteristics to a useful degree may be employed without departing from the broad spirit of my invention; but I prefer to use a condenser, since in these respects it excels any other storage device of which I have knowledge.

In Fig. 2 a modified arrangement of apparatus is illustrated which is particularly adapted for the investigation and utilization of very feeble impulses or disturbances, such as may be used in conveying signals or producing other desired effects at very great distances. In this case the energy stored in the condenser is passed through the primary of a transformer, the secondary circuit of which contains the receiver, and in order to render the apparatus still more suitable for use in detecting feeble impulses, in addition to the sensitive device which is acted upon by the impulses, another such device is included in the secondary circuit of the transformer. The scheme of connections is in the main that of a Wheatstone bridge, the four branches of which are formed by the sensitive device *a* and resistances L, L', and L'', all of which should be preferably inductive and also adjustable in a continuous manner, or at least by very small steps. The condenser C', which is generally made of considerable capacity, is connected to two opposite points of the bridge, while a battery B, in series with a continuously-adjustable non-inductive resistance *r'*, is connected to the other pair of opposite points, as usual. The four resistances included in the branches of the bridge—namely, *a*, L, L', and L''—are of a suitable size and so proportioned that under normal conditions—that is, when the device *a* is not influenced at all or only slightly by the disturbances—there will be no difference of potential, or, in any case, the minimum of the same at the terminals T and T' of the condenser. It is assumed in the present instance that the disturbances to be investigated or utilized are such as will produce a difference of electric potential, however small, between two points or regions in the natural media, as the earth, the water, or the air, and in order to apply this potential difference effectively to the sensitive device *a* the terminals of the same are connected to two plates P and P', which should be of as large a surface as practicable and so located in the media that the largest possible difference of potential will be produced by the disturbances between the terminals of the sensitive device. This device is in the present case one of familiar construction, consisting of an insulating-tube, which is indicated by the heavy lines in the drawings and which has its ends closed tightly by two conducting-plugs with reduced extensions, upon which bear two brushes *b b*, through which the currents are conveyed to the device. The tubular space between the plugs is partially filled with a conducting sensitive powder, as indicated, the proper amount of the same and the size of its grains being determined and adjusted beforehand by experiment. This tube I rotate by clockwork or other means at a uniform and suitable rate of speed, and under these conditions I find that this device behaves toward disturbances of the kind before assumed in a manner similar to that of a stationary cell of selenium toward rays of light. Its electrical resistance is diminished when it is acted upon by the disturbances and is automatically restored upon the cessation of their influence. It is of advantage to employ round grains of powder in the tube, and in any event it is important that they should be of as uniform size and shape as possible and that provision should be made for maintaining an unchanging and very dry atmosphere in the tube. To the terminals T and T' of the condenser C' is connected a coil *p*, usually consisting of a few turns of a conductor of very small resistance, which is the primary of the transformer before referred to, in series with a device *d*, which effects the discharge of the condenser through the coil *p* at predetermined intervals of time. In the present case this device consists of a cylinder made partly of conducting and partly of insulating material *e* and *e'*, respectively, which is rotated at the desired rate of speed by any suitable means. The conducting part *e* is in good electrical connection with shaft S and is provided with tapering segments, as *f*, upon which slides a brush *k*, which should preferably be capable of longitudinal adjustment along the cylinder. Another brush *b'*, which is connected to the condenser-terminal T', being arranged to bear upon the shaft S, it will be seen that whenever the brush *k* comes in contact with a conducting-segment *e* the circuit including the primary *p* will be completed and the condenser, if energized, discharged through the same. By an adjustment of the speed of rotation of the cylinder and a displacement of the brush *k* along the axis of the same the circuit may be made to open and close in as rapid succession and remain open or closed during such intervals of time as may be desired. In inductive relation to the primary *p* is a secondary coil *s*, usually of much thinner wire and of many more turns than the former, to which are connected in a series a receiver R, illustrated as an ordinary magnetic relay, a continuously-adjustable non-inductive resistance *r''*, a battery B' of a properly-determined and very constant electromotive force, and finally a sensitive device *a'* of the same or similar construction as *a*, which is likewise rotated at a uniform speed and which, with its brushes *b' b''*, closes the secondary circuit. The electromotive force of the battery B' is so graduated by means of the adjustable resistance *r''* that the dielectric layers in the sensitive device *a'* are strained very nearly to the point of breaking down and give way upon a slight increase of the electrical pressure on the terminals of the device. It will of course be understood that the resistance *r''* is used mainly because of convenience and that it may be dispensed with, in which case the adjustment may be effected in many other ways, as by determining the proper amount or coarseness of the sensitive powder or by varying the distance apart of the metallic plugs in the ends of the tube. The same may be said of the resistance *r'*, which is in series with the battery B and serves to graduate the force of the latter, so that the dielectric layers of the sensitive device *a* are subjected to a similar strain and maintained in a state of delicate poise.

The various instruments being connected and adjusted in the manner described, it will now be readily seen from the foregoing that under normal conditions, the device *a* being unaffected by the disturbances, or practically so, and there being no or only a very insignificant amount of energy stored in the condenser, the periodical closure of the primary circuit of the transformer through the operation of the device *d* will have no appreciable effect upon the primary coil *p*, and hence no currents will be generated in the secondary coil *s*, at least not such as would disturb the state of delicate balance existing in the secondary circuit including the receiver, and therefore the latter will not be actuated by the battery B'; but when, owing to the disturbances or impulses propagated through the media from a distant source, an additional electromotive force, however small, is created between the terminals of the device *a* the dielectric layers in the same, unable to support the increased strain, give way and allow the current of the battery B to pass through, thus causing a difference of potential at the terminals T and T' of the condenser. A sufficient amount of energy being now stored in this instrument during the time interval between each two succeeding operations of the device *d*, each closure of the primary circuit by the latter results in the passage of a sudden current impulse through the coil *p*, which induces a corresponding current of relatively high electromotive force in the secondary coil *s*. Owing to this the dielectric in the device *a'* gives way, and the current of the battery B' being allowed to pass the receiver R is operated, but only for a moment, since by the rotation of the devices *a*, *a'*, and *d*, which may be all driven from the same shaft, the original conditions are restored, assuming, of course, that the electromotive force set up by the disturbances at the terminals of the sensitive device *a* is only momentary or of a duration not longer than the time of closure of the primary circuit; otherwise the receiver will be actuated a number of times and so long as the influence of the disturbances upon the device *a* continues. In order to render the discharged energy of the condenser more effective in causing the operation of the receiver, the resistance of the primary circuit should be very small and the secondary coil *s* should have a number of turns many times greater than that of the primary coil *p*. It will be noted that since the condenser under the above assumptions is always charged in the same direction the strongest current impulse in the secondary coil, which is induced at the moment when the brush *k* comes in contact with segment *e*, is also of unchanging direction, and for the attainment of the best results it is necessary to connect the secondary coil so that the electromotive force of this impulse will be added to that of the battery and will momentarily strengthen the same. However, under certain conditions, which are well understood by those skilled in the art, the devices will operate whichever way the secondary be connected. It is preferable to make the inductive resistances L and L' relatively large, as they are in a shunt to the device *a* and might, if made too small, impair its sensitiveness. On the other hand, the resistance L'' should not be too large and should be related to the capacity of the condenser and the number of makes and breaks effected by the device *d* in well-known ways. Similar considerations apply, of course, to the circuits including the primary *p* and secondary *s*, respectively.

By carefully observing well-known rules of scientific design and adjustment of the instruments the apparatus may be made extremely sensitive and capable of responding to the feeblest influences, thus making it possible to utilize impulses or disturbances transmitted from very great distances and too feeble to be detected or utilized in any of the ways heretofore known, and on this account the method here described lends itself to many scientific and practical uses of great value.

Obviously the character of the devices and the manner in which they are connected or related may be greatly varied without departing from the spirit of my invention.

What I claim as new, and desire to secure by Letters Patent, is—

1. In an apparatus for utilizing effects or disturbances transmitted through the natural media from a distant source, the combination of an electrical storage device, a charging-circuit connected therewith and including a device sensitive to the action of the effects or disturbances and determining under their control the flow of current in the charging-circuit, a receiving-circuit including a receiver, and means for periodically discharging the storage device through the receiving-circuit, substantially as described.

2. In an apparatus for utilizing effects or disturbances transmitted through the natural media from a distant source, the combination of a condenser, a charging-circuit connected therewith and including a source of electricity and a device sensitive to the action of the effects or disturbances and determining under their control the flow of current in the charging-circuit, a receiving-circuit including a receiver, and means for periodically discharging the condenser through the receiving-circuit, substantially as described.

3. In an apparatus for utilizing effects or disturbances transmitted through the natural media from a distant source, the combination of a circuit including a source of electricity, a storage device adapted to be charged thereby and a device normally of very high resistance but adapted to have its resistance reduced when acted upon by the effects or disturbances, with a receiving-circuit connected with the storage device and including a receiver and a device adapted to open and close the receiving-circuit at predetermined intervals of time, substantially as described.

4. In an apparatus for utilizing effects or disturbances transmitted through the natural media from a distant source, the combination of a circuit including a source of electricity, a condenser and a device normally of very high resistance but adapted to have its resistance reduced when acted upon by the effects or disturbances, with a receiving-circuit connected with the condenser and including a receiver and a device adapted to open and close the receiving-circuit at predetermined intervals of time, substantially as described.

5. In an apparatus for utilizing effects or disturbances transmitted from a distant source, the combination of a circuit including a source of electricity, a storage device adapted to be charged thereby and a device, normally of very high resistance but adapted to have its resistance reduced when acted upon by the effects or disturbances, with a receiving-circuit connected with the storage device and including the primary of a transformer and a device adapted to open and close such second circuit at predetermined intervals of time, and a receiver included in the secondary of the transformer, substantially as described.

6. In an apparatus for utilizing effects or disturbances transmitted from a distant source, the combination of an electrical storage device, a charging-circuit connected therewith and including a device sensitive to the

action of the effects or disturbances and determining under their control the flow of the current in the charging-circuit, and a receiving-circuit supplied with energy from the storage device and including a receiver and a device sensitive to electrical variations in the receiving-circuit, substantially as described.

7. In an apparatus for utilizing effects or disturbances transmitted through the natural media from a distant source, the combination of a condenser, a charging-circuit connected therewith and including a device sensitive to the action of the effects or disturbances and determining under their control the flow of the current in the charging-circuit, and a receiving-circuit supplied with energy from the condenser and including a receiver and a device sensitive to electrical variations in the receiving-circuit, substantially as described.

8. In an apparatus for utilizing effects or disturbances transmitted through the natural media from a distant source, the combination of a circuit, an independent local source of electricity included therein, a storage device connected with the said circuit and adapted to receive energy from the said source, a device normally of very high resistance, but adapted to have its resistance reduced when acted upon by the effects or disturbances, a receiving-circuit connected with the storage device, a transformer, the primary of which is included in said receiving-circuit, a device adapted to open and close the receiving-circuit at predetermined intervals of time, a receiver, and a device, normally of very high resistance, but adapted to have its resistance reduced when acted upon by the effects or disturbances, and included in the secondary circuit of the transformer, as set forth.

9. In an apparatus for utilizing effects or disturbances transmitted from a distant source, the combination with a storage device and an independent source of energy for charging the same, of a receiving-circuit connected with the storage device, a device sensitive to the effects or disturbances and determining under their control the flow of current in the receiving-circuit, substantially as set forth.

10. In an apparatus for utilizing effects or disturbances transmitted through the natural media from a distant source, the combination with a storage device and an independent source of energy for charging the same, of a receiving-circuit connected with the storage device, a device sensitive to the effects or disturbances and adapted to have its resistance varied by the action thereon of such effects or disturbances and determining under their control the flow of current in the receiving-circuit, substantially as set forth.

11. In an apparatus for utilizing effects or disturbances transmitted from a distant source, the combination of a storage device, a battery or similar independent source of energy for charging the same, a sensitive device adapted to have its resistance varied under the influence of the effects or disturbances, a receiver adapted to be operated by the discharge of the storage device, the sensitive device being in one branch of a Wheatstone bridge, the storage device in one of the cross connections between two opposite points of the bridge, and the battery in the other cross connection, and resistances L, L′ and L″ in the three remaining branches of the bridge, as set forth.

12. In an apparatus for utilizing effects or disturbances transmitted through the natural media from a distant source, the combination of a storage device, a battery or like independent source of energy for charging the same, a sensitive device adapted to have its resistance varied under the influence of the effects or disturbances, a circuit connected with the terminals of the storage device, a transformer having its primary in said circuit and a receiver in the secondary circuit of the transformer, the sensitive device being in one branch of a Wheatstone bridge, the storage device in one of the cross connections between opposite points of the bridge and the battery in the other cross connection, and resistances L, L′ and L″ in the three remaining branches of the bridge, as set forth.

NIKOLA TESLA.

Witnesses:
E. A. SUNDERLIN,
D. D. LORD.

The AND logic-gate patents

No. 723,188. PATENTED MAR. 17, 1903.

N. TESLA.

METHOD OF SIGNALING.

APPLICATION FILED JUNE 14, 1901.

NO MODEL.

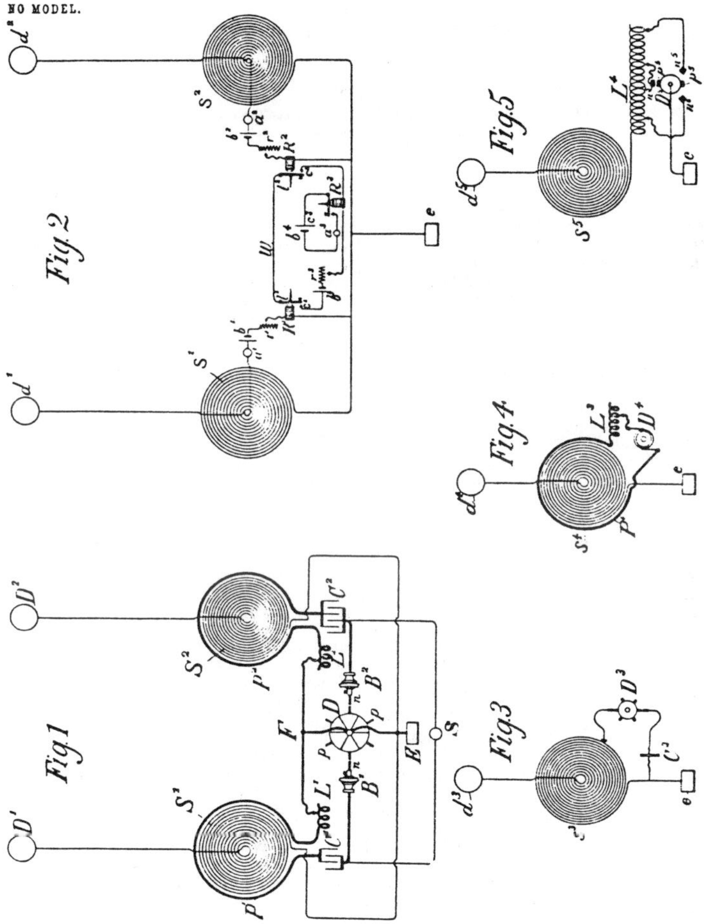

WITNESSES:

Benjamin Miller.

INVENTOR.

Nikola Tesla

BY

Kerr, Page & Cooper

ATTORNEYS.

UNITED STATES PATENT OFFICE.

NIKOLA TESLA, OF NEW YORK, N. Y.

METHOD OF SIGNALING.

SPECIFICATION forming part of Letters Patent No. 723,188, dated March 17, 1903.

Original application filed July 16, 1900, Serial No. 23,847. Divided and this application filed June 14, 1901. Serial No. 64,522. (No model.)

To all whom it may concern:

Be it known that I, NIKOLA TESLA, a citizen of the United States, residing in the borough of Manhattan, in the city, county, and State of New York, have invented certain new and useful Improvements in Methods of Signaling, of which the following is a specification, reference being had to the drawings accompanying and forming a part of the same.

In certain systems for transmitting intelligible messages or governing the movements and operations of distant automata electrical impulses or disturbances produced by suitable apparatus are conveyed through the natural media to a receiving-circuit capable of responding to the impulses, and thereby effecting the control of other appliances. Generally a special device, highly sensitive, is connected to the receiving-circuit, which in order to render it still more susceptible and to reduce the liability of its being affected by extraneous disturbances is carefully adjusted so as to be in tune with the transmitter. By a scientific design of the sending and receiving circuits and other apparatus and skilful adjustment of the same these objects may be in a measure attained; but in long experience I have found that notwithstanding all constructive advantages and experimental resources this method is in many cases inadequate. Thus while I have succeeded in so operating selectively under certain favorable conditions more than one hundred receivers in most cases it is practicable to work successfully but a few, the number rapidly diminishing as, either owing to great distance or other causes, the energy available in the tuned circuits becomes smaller and the receivers necessarily more delicate. Evidently a circuit however well constructed and adjusted to respond exclusively to vibrations of one period is apt to be affected by higher harmonics and still more so by lower ones. When the oscillations are of a very high frequency, the number of the effective harmonics may be large and the receiver consequently easily disturbed by extraneous influences, to such an extent that when very short waves, such as those produced by Hertzian spark apparatus, are used little advantage in this respect is to be derived from tuning the circuits. It being an imperative requirement in most practical applications of such systems of signaling or intelligence transmission that the signals or messages should be exclusive or private, it is highly desirable to do away with the above limitations, especially in view of the fact which I have observed that the influence of powerful electrical disturbances upon sensitive receivers extends even on land to distances of many hundreds of miles, and consequently, in accordance with theory, still farther on sea. To overcome these drawbacks and to enable a great number of transmitting and receiving stations to be operated selectively and exclusively and without any danger of the signals or messages being disturbed, intercepted, or interfered with in any way is the object of my present invention.

Broadly stated, this invention consists in generating two or more kinds or classes of disturbances or impulses of distinctive character with respect to their effect upon a receiving-circuit and operating thereby a distant receiver which comprises two or more circuits, each of which is tuned to respond exclusively to the disturbances or impulses of one kind or class and so arranged that the operation of the receiver is dependent upon their conjoint or resultant action.

By employing only two kinds of disturbances or series of impulses instead of one, as has heretofore been done, to operate a receiver of this kind I have found that safety against the disturbing influences of other sources is increased to such an extent that I believe this number to be amply sufficient in most cases for rendering the exchange of signals or messages reliable and exclusive; but in exceptional instances a greater number may be used and a degree of safety against mutual and extraneous interference attained, such as is comparable to that afforded by a combination-lock. The liability of a receiver being affected by disturbances emanating from other sources, as well as that of the signals or messages being received by instruments for which they are not intended, may, however, be reduced not only by an increased number of the coöperative disturbances or series of impulses, but also by judicious choice of the same and order in which they are made to act upon the receiver.

Evidently there are a great many ways of generating impulses or disturbances at any wave length, wave form, number or order of succession, or of any special character, such as will be capable of fulfilling the requirements above stated, and there are also many ways in which such impulses or disturbances may be made to coöperate and to cause the receiver to be actuated, and inasmuch as the skill and practical knowledge in these novel fields can only be acquired by long experience the degree of safety and perfection attained will necessarily depend upon the ability and resource of the expert who applies my invention; but in order to enable the same to be successfully practiced by any person possessed only of the more general knowledge and experience in these branches I shall describe the simplest plan of carrying it out which is at present known to me.

For a better understanding of the subject reference is now made to the accompanying drawings, in which—

Figures 1 and 2 represent diagrammatically an apparatus and circuit connections employed at the sending and receiving stations, respectively, for the practice of my invention; and Figs. 3, 4, and 5, modified means which may be employed in the practical application of the invention.

In Fig. 1, S' S^2 are two spirally-wound coils or conductors connected with their inner ends to preferably elevated terminals D' and D^2, respectively, and with their outer ends to an earth-plate E. These two coils, conductors, or systems D' S' E and D^2 S^2 E have different and suitably-chosen periods of vibration, and, as pointed out in other patents relating to my system of energy and intelligence transmission, their lengths should be such that the points of maximum pressure developed therein coincide with the elevated terminals D' D^2. By suitably-chosen periods of vibration such periods are meant as will secure the greatest safety against interference, both mutual and extraneous. The two systems may have electrical oscillations impressed upon them in any desired manner conveniently by energizing them through primaries P' and P^2, placed in proximity to them. Adjustable inductances L' and L^2 are preferably included in the primary circuits chiefly for the purpose of regulating the rates of the primary oscillations. In the drawings these primaries P' and P^2 surround the coils S' S^2 and are joined in series through the inductances L' L^2, conductor F, condensers C' and C^2, brush-holders B' and B^2, and a toothed disk D, which is connected to the conductor F and, if desired, also to the ground-plate E, as shown, two independent primary circuits being thus formed. The condensers C' and C^2 are of such capacity and the inductances L' L^2 are so adjusted that each primary is in close resonance with its secondary system, as I have explained in other patents granted to me. The brush-holders B' and B^2 are capable independently of angular and, if necessary, also of lateral adjustment, so that any desired order of succession or any difference of time interval between the discharges occurring in the two primary circuits may be obtained. The condensers being energized from a suitable source S, preferably of high potential, and the disk D being rotated, its projections or teeth *p p* coming at periodically-recurring intervals in very close proximity to or, as the case may be, in contact with conducting rods or brushes *n n* cause the condensers to be discharged in rapid succession through their respective circuits. In this manner the two secondary systems D' S' E and D^2 S^2 E are set in vibration and oscillate freely each at its proper rate for a certain period of time at every discharge. The two vibrations are impressed upon the ground through the plate E and spread to a distance reaching the receiving-station, which has two similar circuits or systems *e s' d'* and *e* s^2 d^2, arranged and connected in the same manner and tuned to the systems at the sending-station, so that each responds exclusively to one of the two vibrations produced by the transmitting apparatus. The same rules of adjustment are observed with respect to the receiving-circuits, care being furthermore taken that the tuning is effected when all the apparatus is connected to the circuits and placed in position, as any change may more or less modify the vibration. Each of the receiving-coils *s'* and s^2 is shunted by a local circuit containing, respectively, sensitive devices *a'* a^2, batteries *b'* b^2, adjustable resistances *r'* r^2, and sensitive relays R' R^2, all joined in series, as shown. The precise connections and arrangements of the various receiving instruments are largely immaterial and may be varied in many ways. The sensitive devices *a'* a^2 may be any of the well-known devices of this kind—as, for example, two conducting-terminals separated by a minute air-gap or a thin film of dielectric which is strained or weakened by a battery or other means to the point of breaking down and gives way to the slightest disturbing influence. Its return to the normal sensitive state may be secured by momentarily interrupting the battery-circuits after each operation or otherwise. The relays R' R^2 have armatures *l'* l^2, which are connected by a wire *w* and when attracted establish electrical contacts at *c'* and c^2, thus closing a circuit containing a battery b^3 and adjustable resistance r^3 and a relay R^3. From the above description it will be readily seen that the relay R^3 will be operated only when both contacts *c'* and c^2 are closed.

The apparatus at the sending-station may be controlled in any suitable manner—as, for instance, by momentarily closing the circuit of the source S, two different electrical vibrations being emitted simultaneously or in rapid succession, as may be desired, at each closure of the circuit. The two receiving-circuits at the distant station, each tuned to respond to the vibrations produced by one of the elements of the transmitter, affect the sensitive devices a' and a^2 and cause the relays R' and R^2 to be operated and contacts c' and c^2 to be closed, thus actuating the receiver or relay R^3, which in turn establishes a contact c^3 and brings into action a device a^3 by means of a battery d', included in a local circuit, as shown. But evidently if through any extraneous disturbance only one of the circuits at the receiving-station is affected the relay R^3 will fail to respond. In this way a communication may be carried on with greatly-increased safety against interference and privacy of the messages may be secured. The receiving-station shown in Fig. 2 is supposed to be one requiring no return message; but if the use of the system is such that this is necessary then the two stations will be similarly equipped, and any well-known means, which it is not thought necessary to illustrate here, may be resorted to for enabling the apparatus at each station to be used in turn as transmitter and receiver. In like manner the operation of a receiver, as R^3, may be made dependent instead of upon two upon more than two such transmitting systems or circuits, and thus any desired degree of exclusiveness or privacy and safety against extraneous disturbances may be attained. The apparatus as illustrated in Figs. 1 and 2 permits, however, special results to be secured by the adjustment of the order of succession of the discharges of the primary circuits P' and P^2 or of the time intervals between such discharges. To illustrate: The action of the relays R' R^2 may be regulated either by adjusting the weights of the levers l' l^2, or the strength of the batteries b' b^2, or the resistances r' r^2, or in other well-known ways, so that when a certain order of succession or time interval between the discharges of the primary circuits P' and P^2 exists at the sending-station the levers l' and l^2 will close the contacts c' and c^2 at the same instant, and thus operate the relay R^3, but will fail to produce this result when the order of succession of or the time interval between the discharges in the primary circuits is another one. By these or similar means additional safety against disturbances from other sources may be attained and, on the other hand, the possibility afforded of effecting the operation of signaling by varying the order of succession of the discharges of the two circuits. Instead of closing and opening the circuit of the source S', as before indicated, for the purpose of sending distinct signals it may be convenient to merely alter the period of either of the transmitting-circuits arbitrarily, as by varying the inductance of the primaries.

Obviously there is no necessity for using transmitters with two or more distinct elements or circuits, as S' and S^2, since a succession of waves or impulses of different characteristics may be produced by an instrument having but one such circuit. A few of the many ways which will readily suggest themselves to the expert who applies my invention are illustrated in Figs. 3, 4, and 5. In Fig. 3 a transmitting system $e\ s^3\ d^3$ is partly shunted by a rotating wheel or disk D^3, which may be similar to that illustrated in Fig. 1 and which cuts out periodically a portion of the coil or conductor s^3, or, if desired, bridges it by an adjustable condenser C^3, thus altering the vibration of the system $e\ s^3\ d^3$ at suitable intervals and causing two distinct kinds or classes of impulses to be emitted in rapid succession by the sender. In Fig. 4 a similar result is produced in the system $e\ s^4\ d^4$ by periodically short-circuiting, through an induction-coil L^3 and a rotating disk D^4 with insulating and conducting segments, a circuit p^4 in inductive relation to said system. Again, in Fig. 5 three distinct vibrations are caused to be emitted by a system $e\ s^5\ d^5$, this result being produced by inserting periodically a number of turns of an induction-coil L^4 in series with the oscillating system by means of a rotating disk B^5 with two projections p^5 p^5 and three rods or brushes n^5, placed at an angle of one hundred and twenty degrees relatively to each other. The three transmitting systems or circuits thus produced may be energized in the same manner as those of Fig. 1 or in any other convenient way. Corresponding to each of these cases the receiving-station may be provided with two or three circuits in an analogous manner to that illustrated in Fig. 2, it being understood, of course, that the different vibrations or disturbances emitted by the sender follow in such rapid succession upon each other that they are practically simultaneous so far as the operation of such relays as R' and R^2 is concerned. Evidently, however, it is not necessary to employ two or more receiving-circuits, but a single circuit may be used also at the receiving-station constructed and arranged like the transmitting-circuits or systems illustrated in Figs. 3, 4, and 5, in which case the corresponding disks, as D^3 D^4 D^5, at the sending will be driven in synchonism with those at the receiving stations as far as may be necessary to secure the desired result; but whatever the nature of the specific devices employed it will be seen that the fundamental idea in my invention is the operation of a receiver by the conjoint or resultant effect of two or more circuits each tuned to respond exclusively to waves, impulses, or vibrations of a certain kind or class produced either simultaneously or successively by a suitable transmitter.

It will be seen from a consideration of the nature of the method hereinbefore described that the invention is applicable not only in the special manner described, in which the transmission of the impulses is effected through natural media, but for the transmission of energy for any purpose and whatever the medium through which the impulses are conveyed.

What I claim is—

1. The method of operating distant receivers which consists in producing and transmitting a plurality of kinds or classes of electrical impulses or disturbances, actuating by the impulses or disturbances of each kind or class one of a plurality of circuits tuned to respond to impulses of such kind or class and operating or controlling the operation of a receiver by the conjoint action of two or more of said circuits, as set forth.

2. The method of signaling, which consists in producing and transmitting a plurality of kinds or classes of electrical impulses or disturbances, developing by the impulses of each class a current in one of a plurality of receiving-circuits tuned to respond exclusively thereto and controlling by means of the conjoint action of such circuits a local circuit, as set forth.

3. The method of signaling which consists in producing a plurality of series of impulses or disturbances differing from each other in character and order of succession, exciting by the impulses of each series one of a plurality of receiving-circuits tuned to respond exclusively thereto and controlling by the conjoint action of such circuits a local circuit, as set forth.

4. The method of signaling which consists in producing a plurality of series of electrical impulses of different character, varying the time interval between the emission of such impulses, exciting by the impulses of each series one of a plurality of receiving-circuits tuned to respond exclusively thereto and controlling by the conjoint action of such circuits a local circuit, as set forth.

5. The method of transmitting electrical energy for conveying intelligible signals which consists in producing a plurality of electrical impulses of different character, developing by the impulses of each kind a current in one of a plurality of receiving-circuits tuned to respond exclusively thereto, controlling the action or effect of the transmitted impulses upon the receiving-circuits by varying the character of said impulses, and operating or controlling the operation of a receiver by the conjoint action of two or more of said receiving-circuits, as set forth.

6. The method of transmitting electrical energy which consists in producing a plurality of electrical waves or impulses of different periodicities, varying the order of transmission of the waves or impulses forming elements of the signal sent, according as one or another receiving-station is to be communicated with where (proper circuit-closing mechanism being provided at each receiving-station) the transmitted signal will be intelligible at and only at the intended receiving-station.

7. The method of transmitting intelligence, which consists in selecting and associating together in predetermined order of succession two or more electrically-generated impulses of different periodicity, forming elements of signals to be sent, and transmitting such selected impulses with reference to the conjoint action of both or all in the production of a signal at a distant point, substantially as set forth.

8. In a system of telegraphy, wherein signals or messages are sent by the use of a plurality of electrical impulses of different periodicities and in a predetermined order of succession, the method of ascertaining at any particular station the particular signal sent to that station, which consists in the selection, to form a signal, of certain transmitted impulses of different periodicities and of a predetermined order of succession to the exclusion of all others, as set forth.

9. The improvement in the art of transmitting electrical energy which consists in operating or controlling a receiving mechanism by a series or group of electrical impulses of different periodicities and of a predetermined order of sucession.

10. In a system for the transmission of electrical energy, for sending signals or messages to any one of two or more receiving-stations, the method of transmitting the message with reference to the intelligible receipt thereof at the desired station, which consists in the transmission of electrical waves or impulses of different periodicities in varying order of transmittal by a separate order or grouping of transmittal for each receiving-station.

NIKOLA TESLA.

Witnesses:
M. LAWSON DYER,
BENJAMIN MILLER.

No. 725,605. PATENTED APR. 14. 1903.

N. TESLA.

SYSTEM OF SIGNALING.

APPLICATION FILED JULY 16, 1900.

NO MODEL.

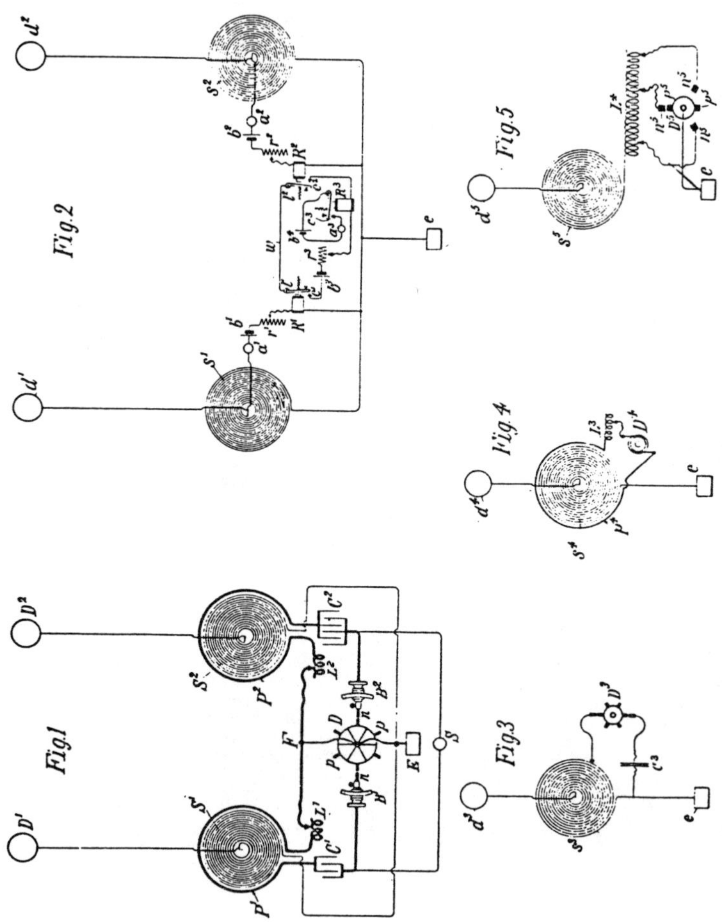

Witnesses:
Raphaël Netter
C. D. Morrill

Nikola Tesla, Inventor
by Kerr, Page & Cooper Attys

UNITED STATES PATENT OFFICE.

NIKOLA TESLA, OF NEW YORK, N.Y.

SYSTEM OF SIGNALING.

SPECIFICATION forming part of Letters Patent No. **725,605**, dated April 14, 1903.

Application filed July 16, 1900. Serial No. 23,847. (No model.)

To all whom it may concern:

Be it known that I, NIKOLA TESLA, a citizen of the United States, residing in the borough of Manhattan, in the city, county, and State of New York, have invented certain new and useful Improvements in Systems of Signaling, of which the following is a specification, reference being had to the drawings accompanying and forming a part of the same.

In certain systems for transmitting intelligible messages or governing the movements and operations of distant automata electrical impulses or disturbances produced by suitable apparatus are conveyed through the natural media to a distant receiving-circuit capable of responding to the impulses, and thereby effecting the control of other appliances. Generally a special device highly sensitive is connected to the receiving-circuit, which in order to render it still more susceptible and to reduce the liability of its being affected by extraneous disturbances is carefully adjusted so as to be in tune with the transmitter. By a scientific design of the sending and receiving circuits and other apparatus and skilful adjustment of the same these objects may be in a measure attained; but in long experience I have found that notwithstanding all constructive advantages and experimental resources this method is in many cases inadequate. Thus while I have succeeded in so operating selectively under certain favorable conditions more than one hundred receivers in most cases it is practicable to work successfully but a few, the number rapidly diminishing as, either owing to great distance or other causes, the energy available in the tune circuits becomes smaller and the receivers necessarily more delicate. Evidently a circuit however well constructed and adjusted to respond exclusively to vibrations of one period is apt to be affected by higher harmonics and still more so by lower ones. When the oscillations are of a very high frequency, the number of the effective harmonics may be large, and the receiver consequently easily disturbed by extraneous influences to such an extent that when very short waves, such as those produced by Hertzian spark apparatus, are used little advantage in this respect is derived from tuning the circuits. It being an imperative requirement in most practical applications of such systems of signaling or intelligence transmission that the signals or messages should be exclusive or private, it is highly desirable to do away with the above limitations, especially in view of the fact, which I have observed, that the influence of powerful electrical disturbances upon sensitive receivers extends, even on land, to distances of many hundreds of miles, and consequently in accordance with theory still farther on sea. To overcome these drawbacks and to enable a great number of transmitting and receiving stations to be operated selectively and exclusively and without any danger of the signals or messages being disturbed, intercepted, or interfered with in any way is the object of my present invention.

Broadly stated, this invention consists in the combination of means for generating and transmitting two or more kinds or classes of disturbances or impulses of distinctive character with respect to their effect upon a receiving-circuit and a distant receiver which comprises two or more circuits of different electrical character or severally tuned, so as to be responsive to the different kinds or classes of impulses and which is dependent for operation upon the conjoint or resultant action of the two or more circuits or the several instrumentalities controlled or operated thereby. By employing only two kinds of disturbances or series of impulses instead of one, as has heretofore been done to operate a receiver of this kind, I have found that safety against the disturbing influences of other sources is increased to such an extent that I believe this number to be amply sufficient in most cases for rendering the exchange of signals or messages reliable and exclusive; but in exceptional instances a greater number may be used and a degree of safety against mutual extraneous interference attained, such as is comparable to that afforded by a combination-lock. The liability of a receiver being affected by disturbances amanating from other sources, as well as that of the signals or messages being received by instruments for which they are not intended, may, however, be reduced not only by an increased number of the coorperative disturbances or series of impulses, but also by a judicious choice of the same and the order in which they are made to act upon the receiver.

Evidently there are a great many ways of generating impulses or disturbances of any wave length, wave form, number or order of succession, or of any special character such as will be capable of fulfilling the requirements above stated, and there are also many ways in which such impulses or disturbances may be made to cooperate and to cause the receiver to be actuated, and inasmuch as the skill and practical knowledge in these novel fields can only be acquired by long experience the degree of safety and perfection attained will necessarily depend upon the ability and resource of the expert who applies my invention; but in order to enable the same to be successfully practiced by any person possessed only of the more general knowledge and experience in these branches I shall describe the simplest plan of carrying it out which is at present known to me.

For a better understanding of the subject reference is now made to the accompanying drawings, in which —

Figures 1 and 2 represent diagrammatically the apparatus and circuit connections employed at the sending and receiving stations, respectively; and Figs. 3, 4, and 5 modified means which may be employed in the practical application of the invention.

In Fig. 1, S' and S^2 are two spirally-wound coils or conductors connected with their inner ends to elevated terminals D' and D^2, respectively, and with their outer ends to an earth-plate E. These two coils, conductors, or systems D' S' E and D^2 S^2 E have different and suitably-chosen periods of vibration, and, as pointed out in other patents relating to my system of energy and intelligence transmission, their lengths should be such that the points of maximum pressure developed therein coincide with the elevated terminals D' D^2. The two systems may have electrical oscillations impressed upon them in any desired manner, conveniently by energizing them through primaries P' and P^2, placed in proximity to them. Adjustable inductances L' and L^2 are preferably included in the primary circuits chiefly for the purpose of regulating the rates of the primary oscillations. In the drawings these primaries P' and P^2 surround the coils S' S^2 and are joined in series through the inductances L' L^2, conductor F, condensers C' and C^2, brush-holders B' and B^2, and a toothed disk D, which is connected to the conductor F and, if desired, also to the ground-plate E, as shown, two independent primary circuits being thus formed. The condensers C' and C^2 are of such capacity and the inductances L' and L^2 are so adjusted that each primary is in close resonance with its secondary system, as I have explained in other patents granted to me. The brush holders B' and B^2 are capable independently of angular and, if necessary, also of lateral adjustment, so that any desired order of succession or any difference of time interval between the discharges occurring in the two primary circuits may be obtained. The condensers being energized from a suitable source S, preferably of high potential, and the disk D being rotated, its projections or teeth *p p*, coming at periodically-recurring intervals in very close proximity to or, as the case may be, in contact with conducting rods or brushes *n n*, cause the condensers to be discharged in rapid succession through their respective circuits. In this matter the two secondary systems D' S' E and D^2 S^2 E are set in vibration and oscillate freely, each at its proper rate, for a certain period of time at every discharge. The two vibrations are impressed upon the ground through the plate E and spread to a distance reaching the receiving station, which has two similar circuits or systems $e\ s'\ d'$ and $e\ s^2\ d^2$ arranged and connected in the same manner and tuned to the systems at the sending-station, so that each responds exclusively to one of the two vibrations produced by the transmitting apparatus. The same rules of adjustment are observed with respect to the receiving-circuits, care being, furthermore, taken that the tuning is effected when all the apparatus is connected to the circuits and placed in position, as any change may more or less modify the vibration. Each of the receiving coils s' and s^2 is shunted by a local circuit containing, respectively, sensitive devices a' a^2, batteries b' b^2, adjustable resistances r' r^2, and sensitive relays R' R^2, all joined in series, as shown. The precise connections and arrangements of the various receiving instruments are largely immaterial and may be varied in many ways. The sensitive devices a' a^2 may be any of the well-known devices of this kind — as, for example, two conducting-terminals separated by a minute air-gap or a thin film of dielectric which is strained or weakened by a battery or other means to the point of breaking down and gives way to the slightest disturbing influence. Its return to the normal sensitive state may be secured by momentarily interrupting the battery-circuit after each operation or otherwise. The relays R' R^2 have armatures l' L^2, which are connected by a wire *w* and when attracted establish electrical contacts c' and c^2, thus closing a circuit containing a battery b^3, and adjustable resistance r^3, and a relay R^3.

From the above description it will be readily seen that the relay R^3 will be operated only when both contacts c' and c^2 are closed.

The apparatus at the sending-station may be controlled in any suitable manner — as, for instance, by momentarily closing the circuit of the source S, two different electric vibrations being emitted simultaneously or in rapid succession, as may be desired, at each closure of the circuit. The two receiving-circuits at the distant station, each tuned to respond to the vibrations produced by one of the elements of the transmitter, affect the sensitive devices a' and a^2 and cause the relays R' and R^2 to be operated and contacts c' and c^2 to be closed, thus actuating the receiver or relay R^3, which in turn establishes a contact c^3 and brings into action a device a^3 by means of a battery d^4, included in a local circuit, as shown; but evidently if through any extraneous disturbance only one of the circuits at the receiving-station is affected the relay R^3 will fail to respond. In this way communication may be carried on with greatly-increased safety against interference and privacy of the messages may be secured. The receiving-station (shown in Fig. 2) is supposed to be one requiring no return message; but if the use of the system is such that this is necessary then the two stations will be similarly equipped and any well-known means, which it is not thought necessary to illustrate here, may be resorted to for enabling the apparatus at each station to be used in turn as transmitter and receiver. In like manner the operation of a receiver, as R^3, may be made dependent, instead of upon two, upon more than two such transmitting systems or circuits, and thus any desired degree of exclusiveness or privacy and safety against extraneous disturbances may be attained. The apparatus as illustrated in Figs. 1 and 2 permits, however, special results to be secured by the adjustment of the order of succession of the discharge of the primary circuits P' and P^2 or of the time interval between such discharges. To illustrated, the action of the relays R' R^2 may be regulated either by adjusting the weights of the levers l' l^2, or the strength of the batteries b' b^2, or in other well-known ways, so that when a certain order of succession or timeinterval between the discharges of the primary circuits P' and P^2 exists at the sending-station the levers l' and l^2 will close the contacts c' and c^2 at the same instant, and thus operate the relay R^3; but it will fail to produce this result when the order of succession of or the time interval between the discharges in the primary circuits is another one. By these or similar means additional safety against disturbances from other sources may be attained and, on the other hand, the possibility afforded of effecting the operation of signaling by varying the order of succession of the discharges of the two circuits. Instead of closing and opening the circuit of the source S, as before indicated, for the purpose of sending distinct signals it may be convenient to merely alter the period of either of the transmitting-circuits arbitrarily, as by varying the inductance of the primaries.

Obviously there is no necessity for using transmitters with two or more distinct elements or circuits, as S' and S^2, since a succession of waves or impulses of different characteristics may be produced by an instrument having but one such circuit. A few of the many ways which will readily suggest themselves to the expert who applies my invention are illustrated in Figs. 3, 4, and 5. In Fig. 3 a transmitting system $e\ s^3\ d^3$ is partly shunted by a rotating wheel or disk D^3, which may be similar to that illustrated in Fig. 1 and which cuts out periodically a portion of the coil or conductor s^3 or, if desired, bridges it by an adjustable condenser C^3, thus altering the vibration of the system $e\ s^3\ d^3$ at suitable intervals and causing two distinct kinds or classes of impulses to be emitted in rapid succession by the sender. In Fig. 4 a similar result is produced in the system $e\ s^4\ d^4$ by periodically short-circuiting, through an induction-coil L^3 and a rotating disk D^4 with insulating and conducting segments, a circuit p^4 in inductive relation to said system. Again, in Fig. 5 three distinct vibrations are caused to be emitted by a system $e\ s^5\ d^5$, this result being produced by inserting periodically a suitable number of turns of an induction-coil L^4 in series with the oscillating system by means of a rotating disk B^5 with two projections p^5 p^5 and three rods or brushes n^5, placed at an angle of one hundred and twenty degrees relatively to each other. The three transmitting systems or circuits thus produced may be energized in the same manner as those of Fig. 1 or in any other convenient way. Corresponding to each of these cases the receiving-station may be provided with two or three circuits in an analogous manner to that illustrated in Fig. 2, it being understood, of course, that the different vibrations or disturbances emitted by the sender follow in such rapid succession upon each other that they are practically simultaneous, so far as the operation of such relays as R' and R^2 is concerned. Evidently, however, it is not necessary to employ two or more receiving-circuits; but a single circuit may be used also at the receiving-station constructed and arranged like the transmitting circuits or systems illustrated in Figs. 3, 4, and 5, in which case the corresponding disks, as D^3 D^4 D^5, at the sending will be driven in synchronism with those at the receiving stations as far as may be necessary to secure the desired result; but whatever the nature of the specific devices employed it will be seen that the fundamental idea in my invention is the operation of a receiver by the conjoint or resultant effect of two or more circuits each tuned to respond exclusively to waves, impulses, or vibrations of a certain kind or class produced either simultaneously or successively by a suitable transmitter.

It will be seen from a consideration of the nature of the method hereinbefore described that the invention is applicable not only in the special manner described, in which the transmission of the impulses is effected through natural media, but for the transmission of energy for any purpose and whatever the medium through which the impulses are conveyed.

What I claim is —

1. In a system for the transmission of electrical energy, the combination with means for

producing two or more distinctive kinds of disturbances or impulses, of receiving-circuits, each tuned to respond to the waves or impulses of one kind only, and a receiving device dependent for operation upon the conjoint action of the several receiving-circuits, as set forth.

2. In a system for the transmission of electrical impulses and the operation or control, of signaling or other apparatus thereby, the combination with a transmitter adapted to produce two or more distinctive kinds or classes of disturbances or impulses, of sensitive receiving-circuits, each tuned to respond to the impulses or disturbances of one kind or class only, and a receiving device dependent for operation upon the conjoint action of the sensitive circuits, as set forth.

3. In a system for the transmission of electrical impulses, and the operation or control of signaling, or other apparatus thereby, the combination with a transmitter adapted to produce two or more distinctive kinds or classes of disturbances or impulses, of sensitive circuits at the receiving point or station, each tuned to respond to the impulses or disturbances of one kind or class only, a local circuit arranged to be completed by the conjoint action of the sensitive circuits and a receiving device connected therewith, as set forth.

4. In a system for the transmission of electrical impulses, and the operation or control of signaling or other apparatus thereby, the combination with a transmitting apparatus adapted to produce two or more distinctive kinds of disturbances or impulses, of means for varying the time intervals of the emission of the impulses of the several kinds, sensitive circuits each tuned to respond to the impulses or disturbances of one kind only, and a receiving apparatus dependent for operation upon the conjoint action of the sensitive circuits, as set forth.

5. In a system, such as herein described, the combination with a transmitter adapted to produce a plurality of distinctive kinds of electrical disturbances or impulses, of a receiving apparatus comprising a plurality of circuits, a sensitive device and a relay included in each circuit, and each said circuit being tuned to respond to the impulses or disturbances of one kind only, and a receiving apparatus in a local circuit controlled by the relays and adapted to be completed by the conjoint action of all of said relays, as set forth.

6. In a system of the kind described, the combination with a transmitter adapted to produce two or more series of electrical oscillations or impulses of different frquencies, of a receiving apparatus comprising plurality of sensitive circuits each tuned to respond to the impulses of one of the series produced by the transmitter, and a signaling device dependent for its operation upon the conjoint action of said circuits, as set forth.

7. The combination with a plurality of transmitter elements, each adapted to produce a series impulses or disturbances of a distinctive character, and means for controlling and adjusting the same, of a receiver having a plurality of sensitive circuits each tuned so as to be affected by one of the series of impulses only, and dependent for operation upon the conjoint action of all of said circuits, as set forth.

8. The combination with a transmitter adapted to produce series of electrical impulses or disturbances of distinctive character and in a given order of succession, of a receiving apparatus comprising tuned circuits responding to such impulses in a corresponding order, and dependent for operation upon the conjoint action of said elements, as set forth.

9. In A receiving apparatus, the combination with a plurality of sensitive circuits, severally tuned to respond to waves or impulses of a different kind or class a receiving circuit controlled by the sensitive circuits and a device connected with the receiving-circuits adapted to be operated when said circuit is completed by the conjoint action of two or more of the sensitive circuits, as set forth.

10. A system for the transmission of electrical energy, having in combination means for producing and transmitting two or more impulses of different periodicities to form a signal in a predetermined order of succession, as set forth.

11. In a system for the transmission of electrical energy, the combination with a transmitting apparatus comprising one or more circuits, means for impressing therein oscillations or impulses of different character and a receiving apparatus comprising a plurality of circuits each tuned to respond to the impulses of one kind produced by the transmitter and a receiver dependent for operation upon the conjoint action of the receiving-circuits, as set forth.

12. In a system for the transmission of electrical energy, the combination with a transmitting apparatus comprising a transformer and means for impressing upon the secondary element of the same oscillations or impulses of different character, of a receiving apparatus comprising a plurality of circuits each tuned to the impulses of one kind emitted by the secondary of the transitting-transformer, and a receiver dependent for operation upon the conjoint action of the receiving-circuits, as set forth.

13. In a system for the transmission of electrical energy, the combination with a transmitting apparatus comprising a transformer and means for impressing upon the secondary elements of the same oscillations or impulses of different periodicities and in a given order of succession, of a receiving apparatus comprising a plurality of circuits each tuned to respond to the transmitted impulses of one period, and a receiver dependent for operation upon the conjoint action of the receiving-circuits, as set forth.

14. In a signaling system, the combination of

725,605 5

means for generating a series of electrical impulses of different periodicities, receiving-circuits of differing electrical periods of vibration, and an indicating mechanism operated to give an intelligible indication only when currents are induced in the receiving-circuits a predetermined order, as set forth.

15. In a system for the transmission of energy, the combination of two or more circuits differing with respect of one of their electrical constants, means for energizing said circuits, and an indicating mechanism operative only by conjoint action of two or more currents generated by waves from the sending-station, as set forth.

16. In a system for the transmission of electrical energy, the combination with a transmitter adapted to produce electrical waves or oscillations varying in character in a predetermined order, of a receiving instrument responsive to said oscillations and dependent for operation upon the action thereof in a corresponding order, as set forth.

NIKOLA TESLA.

Witnesses:
John C. Kerr,
Richard S. Donovan.

The Tesla AND logic gate patents

The pair of patents Nos. 723,188 and 725,605, are 'method' and 'means,' respectively of the invention.

Differences in descriptions of the invention in the method and means patents are slight. Aside from the use of excessive punctuation in the former, an interesting addition occurs at line 15 of the latter: "a *distant* receiving circuit." The capability to transmit and receive signals at great distances had become a dominant issue with Tesla since the alleged transmission of the letter *S* across the Atlantic by Marconi between the dates of application for the method and means patents.

The other differences occur in lines 72-82 of the means patent, which language reflects apparatus for accomplishing the invention, and a greater number of claims. In only a few years, the means of accomplishing the invention disclosed by Tesla became passé by advances in technologies.

No. 787,412. PATENTED APR. 18, 1905.

N. TESLA.

ART OF TRANSMITTING ELECTRICAL ENERGY THROUGH THE NATURAL MEDIUMS.

APPLICATION FILED MAY 16, 1900. RENEWED JUNE 17, 1902.

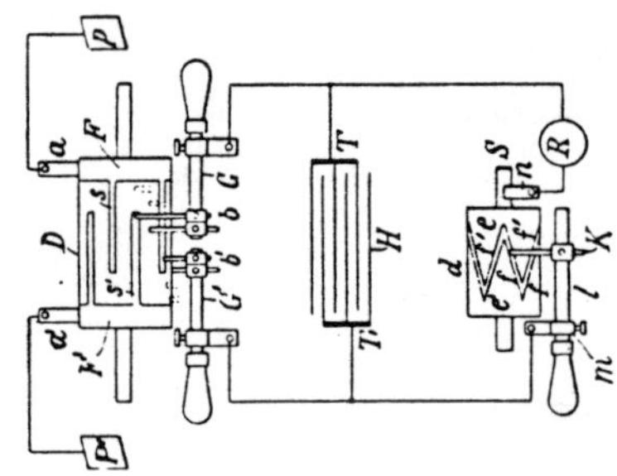

Fig. 2

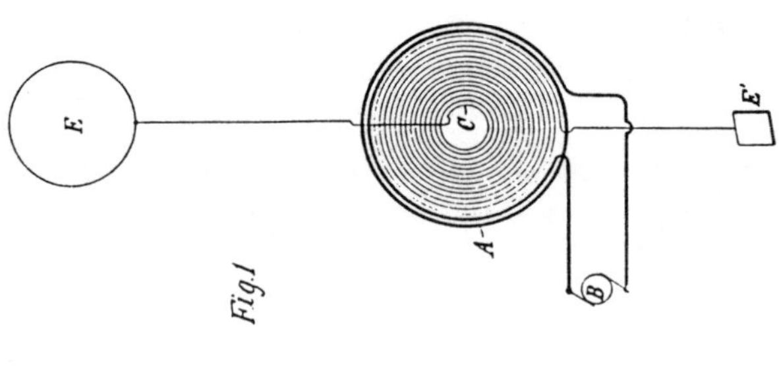

Fig. 1

Witnesses:
Raphaël Netter
M. Lawson Dyer.

Nikola Tesla Inventor
by Kerr, Page & Cooper Attys

No. 787,412. Patented April 18, 1905.

UNITED STATES PATENT OFFICE.

NIKOLA TESLA, OF NEW YORK, N. Y.

ART OF TRANSMITTING ELECTRICAL ENERGY THROUGH THE NATURAL MEDIUMS.

SPECIFICATION forming part of Letters Patent No. 787,412, dated April 18, 1905.

Application filed May 16, 1900. Renewed June 17, 1902. Serial No. 112,034.

To all whom it may concern:

Be it known that I, NIKOLA TESLA, a citizen of the United States, residing in the borough of Manhattan, in the city, county, and State of New York, have discovered a new and useful Improvement in the Art of Transmitting Electrical Energy Through the Natural Media, of which the following is a specification, reference being had to the drawings accompanying and forming a part of the same.

It is known since a long time that electric currents may be propagated through the earth, and this knowledge has been utilized in many ways in the transmission of signals and the operation of a variety of receiving devices remote from the source of energy, mainly with the object of dispensing with a return conducting-wire. It is also known that electrical disturbances may be transmitted through portions of the earth by grounding only one of the poles of the source, and this fact I have made use of in systems which I have devised for the purposes of transmitting through the natural media intelligible signals or power and which are now familiar; but all experiments and observations heretofore made have tended to confirm the opinion held by the majority of scientific men that the earth, owing to its immense extent, although possessing conducting properties, does not behave in the manner of a conductor of limited dimensions with respect to the disturbances produced, but, on the contrary, much like a vast reservoir or ocean, which while it may be locally disturbed by a commotion of some kind remains unresponsive and quiescent in a large part or as a whole. Still another fact now of common knowledge is that when electrical waves or oscillations are impressed upon such a conducting-path as a metallic wire reflection takes place under certain conditions from the ends of the wire, and in consequence of the interference of the impressed and reflected oscillations the phenomenon of "stationary waves" with maxima and minima in definite fixed positions is produced. In any case the existence of these waves indicates that some of the outgoing waves have reached the boundaries of the conducting-path and have been reflected from the same. Now I have discovered that notwithstanding its vast dimensions and contrary to all observations heretofore made the terrestrial globe may in a large part or as a whole behave toward disturbances impressed upon it in the same manner as a conductor of limited size, this fact being demonstrated by novel phenomena, which I shall hereinafter describe.

In the course of certain investigations which I carried on for the purpose of studying the effects of lightning discharges upon the electrical condition of the earth I observed that sensitive receiving instruments arranged so as to be capable of responding to electrical disturbances created by the discharges at times failed to respond when they should have done so, and upon inquiring into the causes of this unexpected behavior I discovered it to be due to the character of the electrical waves which were produced in the earth by the lightning discharges and which had nodal regions following at definite distances the shifting source of the disturbances. From data obtained in a large number of observations of the maxima and minima of these waves I found their length to vary approximately from twenty-five to seventy kilometers, and these results and certain theoretical deductions led me to the conclusion that waves of this kind may be propagated in all directions over the globe and that they may be of still more widely differing lengths, the extreme limits being imposed by the physical dimensions and properties of the earth. Recognizing in the existence of these waves an unmistakable evidence that the disturbances created had been conducted from their origin to the most remote portions of the globe and had been thence reflected, I conceived the idea of producing such waves in the earth by artificial means with the object of utilizing them for many useful purposes for which they are or might be found applicable. This problem was rendered extremely difficult owing to the immense dimensions of the planet, and consequently enormous movement of electricity or rate at which electrical energy had to be delivered in order to approximate, even in a remote degree, movements or rates which are manifestly attained in the displays of electrical forces in nature and which seemed at first unrealizable by any human agencies; but by gradual and continuous improvements of a generator of electrical oscillations, which I have described in my Patents Nos. 645,576 and 649,621, I finally succeeded in reaching electrical movements or rates of delivery of electrical energy not only approximating, but, as shown in many comparative tests and measurements, actually surpassing those of lightning discharges, and by means of this apparatus I have found it possible to reproduce whenever desired phenomena in the earth the same as or similar to those due to such discharges. With the knowledge of the phenomena discovered by me and the means at command for accomplishing these results I am enabled not only to carry out many operations by the use of known instruments, but also to offer a solution for many important problems involving the operation or control of remote devices which for want of this knowledge and the absence of these means have heretofore been entirely impossible. For example, by the use of such a generator of stationary waves and receiving apparatus properly placed and adjusted in any other locality, however remote, it is practicable to transmit intelligible signals or to control or actuate at will any one or all of such apparatus for many other important and valuable purposes, as for indicating wherever desired the correct time of an observatory or for ascertaining the relative position of a body or distance of the same with reference to a given point or for determining the course of a moving object, such as a vessel at sea, the distance traversed by the same or its speed, or for producing many other useful effects at a distance dependent on the intensity, wave length, direction or velocity of movement, or other feature or property of disturbances of this character.

I shall typically illustrate the manner of applying my discovery by describing one of the specific uses of the same—namely, the transmission of intelligible signals or messages between distant points—and with this object reference is now made to the accompanying drawings, in which—

Figure 1 represents diagrammatically the generator which produces stationary waves in the earth, and Fig. 2 an apparatus situated in a remote locality for recording the effects of these waves.

In Fig. 1, A designates a primary coil forming part of a transformer and consisting generally of a few turns of a stout cable of inappreciable resistance, the ends of which are connected to the terminals of a source of powerful electrical oscillations, diagrammatically represented by B. This source is usually a condenser charged to a high potential and discharged in rapid succession through the primary, as in a type of transformer invented by me and not well known; but when it is desired to produce stationary waves of great lengths an alternating dynamo of suitable construction may be used to energize the primary A. C is a spirally-wound secondary coil within the primary having the end nearer to the latter connected to the ground E′ and the other end to an elevated terminal E. The physical constants of coil C, determining its period of vibration, are so chosen and adjusted that the secondary system E′ C E is in the closest possible resonance with the oscillations impressed upon it by the primary A. It is, moreover, of the greatest importance in order to still further enhance the rise of pressure and to increase the electrical movement in the secondary system that its resistance be as small as practicable and its self-induction as large as possible under the conditions imposed. The ground should be made with great care, with the object of reducing its resistance. Instead of being directly grounded, as indicated, the coil C may be joined in series or otherwise to the primary A, in which case the latter will be connected to the plate E′; but be it that none or a part or all of the primary or exciting turns are included in the coil C the total length of the conductor from the ground-plate E′ to the elevated terminal E should be equal to one-quarter of the wave length of the electrical disturbance in the system E′ C E or else equal to that length multiplied by an odd number. This relation being observed, the terminal E will be made to coincide with the points of maximum pressure in the secondary or excited circuit, and the greatest flow of electricity will take place in the same. In order to magnify the electrical movement in the secondary as much as possible, it is essential that its inductive connection with the primary A should not be very intimate, as in ordinary transformers, but loose, so as to permit free oscillation—that is to say, their mutual induction should be small. The spiral form of coil C secures this advantage, while the turns near the primary A are subjected to a strong inductive action and develop a high initial electromotive force. These adjustments and relations being carefully completed and other constructive features indicated rigorously observed, the electrical movement produced in the secondary system by the inductive action of the primary A will be enormously magnified, the increase being directly proportionate to the inductance and frequency and inversely to the resistance of the secondary system. I have found it practicable to produce in this manner an electrical movement thousands of times greater than the initial—that is, the one impressed upon the secondary by the primary A—and I have thus reached activities or rates of flow of electrical energy in the system E′ C E measured by many tens of thousands of horsepower. Such immense movements of electricity give rise to a variety of novel and striking phenomena, among which are those already described. The powerful electrical oscillations in the system E′ C E being communicated to the ground cause corresponding vibrations to be propagated to distant parts of the globe, whence they are reflected and by interference with the outgoing vibrations produce stationary waves the crests and hollows of which lie in parallel circles relatively to which the ground-plate E′ may be considered to be the pole. Stated otherwise, the terrestrial conductor is thrown into resonance with the oscillations impressed upon it just like a wire. More than this, a number of facts ascertained by me clearly show that the movement of electricity through it follows certain laws with nearly mathematical rigor. For the present it will be sufficient to state that the planet behaves like a perfectly smooth or polished conductor of inappreciable resistance with capacity and self induction uniformly distributed along the axis of symmetry of wave propagation and transmitting slow electrical oscillations without sensible distortion and attenuation.

Besides the above three requirements seem to be essential to the establishment of the resonating condition.

First. The earth's diameter passing through the pole should be an odd multiple of the quarter wave length—that is, of the ratio between the velocity of light—and four times the frequency of the currents.

Second. It is necessary to employ oscillations in which the rate of radiation of energy into space in the form of hertzian or electromagnetic waves is very small. To give an idea, I would say that the frequency should be smaller than twenty thousand per second, though shorter waves might be practicable. The lowest frequency would appear to be six per second, in which case there will be but one node, at or near the ground-plate, and, paradoxical as it may seem, the effect will increase with the distance and will be greatest in a region diametrically opposite the transmitter. With oscillations still slower the earth, strictly speaking, will not resonate, but simply act as a capacity, and the variation of potential will be more or less uniform over its entire surface.

Third. The most essential requirement is, however, that irrespective of frequency the wave or wave-train should continue for a certain interval of time, which I have estimated to be not less than one-twelfth or probably 0.08484 of a second and which is taken in passing to and returning from the region diametrically opposite the pole over the earth's surface with a mean velocity of about four hundred and seventy-one thousand two hundred and forty kilometers per second.

The presence of the stationary waves may be detected in many ways. For instance, a circuit may be connected directly or inductively to the ground and to an elevated terminal and tuned to respond more effectively to the oscillations. Another way is to connect a tuned circuit to the ground at two points lying more or less in a meridian passing through the pole E′ or, generally stated, to any two points of a different potential.

In Fig. 2 I have shown a device for detecting the presence of the waves such as I have used in a novel method of magnifying feeble effects which I have described in my Patents Nos. 685,953 and 685,955. It consists of a cylinder D, of insulating material, which is moved at a uniform rate of speed by clockwork or other suitable motive power and is provided with two metal rings F F′, upon which bear brushes *a* and *a*′, connected, respectively, to the terminal plates P and P′. From the rings F F′ extend narrow metallic segments *s* and *s*′, which by the rotation of the cylinder D are brought alternately into contact with double brushes *b* and *b*′, carried by and in contact with conducting-holders *h* and *h*′, supported in metallic bearings G and G′, as shown. The latter are connected to the terminals T and T′ of a condenser H, and it should be understood that they are capable of angular displacement as ordinary brush-supports. The object of using two brushes, as *b* and *b*′, in each of the holders *h* and *h*′ is to vary at will the duration of the electric contact of the plates P and P′ with the terminals T and T′, to which is connected a receiving-circuit including a receiver R and a device *d*, performing the duty of closing the receiving-circuit at predetermined intervals of time and discharging the stored energy through the receiver. In the present case this device consists of a cylinder made partly of conducting and partly of insulating material *e* and *e*′, respectively, which is rotated at the desired rate of speed by any suitable means. The conducting part *e* is in good electrical connection with the shaft S and is provided with tapering segments *f f*′, upon which slides a brush *k*, supported on a conducting-rod *l*, capable of longitudinal adjustment in a metallic support *m*. Another brush, *n*, is arranged to bear upon the shaft S, and it will be seen that whenever one of the segments *f*′ comes in contact with the brush *k* the circuit including the receiver R is completed and the condenser discharged through the same. By an adjustment of the speed or rotation of the cylinder *d* and a displacement of the brush *k* along the cylinder the circuit may be made to open and close in as rapid succession and remain open or closed during such intervals of time as may be desired. The plates P and P′, through which the electrical energy is conveyed to the brushes *a* and *a*′, may be at a considerable distance from each other in the ground or one in the ground and the other in the air, preferably at some height. If but one plate is connected to earth and the other maintained at an elevation, the location of the apparatus must be determined with reference to the position of the stationary waves established by the generator, the effect evidently being greatest in a maximum and zero in a nodal region. On the other hand, if both plates be connected to earth the points of connection must be selected with reference to the difference of potential which it is desired to secure, the strongest effect being of course obtained when the plates are at a distance equal to half the wave length.

In illustration of the operation of the system let it be assumed that alternating electrical impulses from the generator are caused to produce stationary waves in the earth, as above described, and that the receiving apparatus is properly located with reference to the position of the nodal and ventral regions of the waves. The speed of rotation of the cylinder D is varied until it is made to turn in synchronism with the alternate impulses of the generator, and the position of the brushes *b* and *b'* is adjusted by angular displacement or otherwise, so that they are in contact with the segments S and S' during the periods when the impulses are at or near the maximum of their intensity. These requirements being fulfilled, electrical charges of the same sign will be conveyed to each of the terminals of the condenser, and with each fresh impulse it will be charged to a higher potential. The speed of rotation of the cylinder *d* being adjustable at will, the energy of any number of separate impulses may thus be accumulated in potential form and discharged through the receiver R upon the brush *k* coming in contact with one of the segments *f'*. It will be understood that the capacity of the condenser should be such as to allow the storing of a much greater amount of energy than is required for the ordinary operation of the receiver. Since by this method a relatively great amount of energy and in a suitable form may be made available for the operation of a receiver, the latter need not be very sensitive; but when the impulses are very weak or when it is desired to operate a receiver very rapidly any of the well-known sensitive devices capable of responding to very feeble influences may be used in the manner indicated or in other ways. Under the conditions described it is evident that during the continuance of the stationary waves the receiver will be acted upon by current impulses more or less intense, according to its location with reference to the maxima and minima of said waves; but upon interrupting or reducing the flow of the current the stationary waves will disappear or diminish in intensity. Hence a great variety of effects may be produced in a receiver, according to the mode in which the waves are controlled. It is practicable, however, to shift the nodal and ventral regions of the waves at will from the sending-station, as by varying the length of the waves under observance of the above requirements. In this manner the regions of maximum and minimum effect may be made to coincide with any receiving station or stations. By impressing upon the earth two or more oscillations of different wave length a resultant stationary wave may be made to travel slowly over the globe, and thus a great variety of useful effects may be produced. Evidently the course of a vessel may be easily determined without the use of a compass, as by a circuit connected to the earth at two points, for the effect exerted upon the circuit will be greatest when the plates P P' are lying on a meridian passing through ground-plate E' and will be *nil* when the plates are located at a parallel circle. If the nodal and ventral regions are maintained in fixed positions, the speed of a vessel carrying a receiving apparatus may be exactly computed from observations of the maxima and minima regions successively traversed. This will be understood when it is stated that the projections of all the nodes and loops on the earth's diameter passing through the pole or axis of symmetry of the wave movement are all equal. Hence in any region at the surface the wave length can be ascertained from simple rules of geometry. Conversely, knowing the wave length, the distance from the source can be readily calculated. In like ways the distance of one point from another, the latitude and longitude, the hour, &c., may be determined from the observation of such stationary waves. If several such generators of stationary waves, preferably of different length, were installed in judiciously-selected localities, the entire globe could be subdivided in definite zones of electric activity, and such and other important data could be at once obtained by simple calculation or readings from suitably-graduated instruments. Many other useful applications of my discovery will suggest themselves, and in this respect I do not wish to limit myself. Thus the specific plan herein described of producing the stationary waves might be departed from. For example, the circuit which impresses the powerful oscillations upon the earth might be connected to the latter at two points. In this application I have advanced various improvements in means and methods of producing and utilizing electrical effects which either in connection with my present discovery or independently of the same may be usefully applied.

I desire it to be understood that such novel features as are not herein specifically claimed will form the subjects of subsequent applications.

What I now claim is—

1. The improvement in the art of transmitting electrical energy to a distance which consists in establishing stationary electrical waves in the earth, as set forth.

2. The improvement in the art of transmit-

ting electrical energy to a distance which consists in impressing upon the earth electrical oscillations of such character as to produce stationary electrical waves therein, as set forth.

3. The improvement in the art of transmitting and utilizing electrical energy which consists in establishing stationary electrical waves in the natural conducting media, and operating thereby one or more receiving devices remote from the source of energy, as set forth.

4. The improvement in the art of transmitting and utilizing electrical energy which consists in establishing in the natural conducting media, stationary electrical waves of predetermined length and operating thereby one or more receiving devices remote from the source of energy and properly located with respect to the position of such waves, as herein set forth.

5. The improvement in the art of transmitting and utilizing electrical energy, which consists in establishing in the natural conducting media, stationary electrical waves, and varying the length of such waves, as herein set forth.

6. The improvement in the art of transmitting and utilizing electrical energy, which consists in establishing in the natural conducting media stationary electrical waves and shifting the nodal and ventral regions of these waves, as described.

NIKOLA TESLA.

Witnesses:
M. LAWSON DYER,
BENJAMIN MILLER.

"Inductorium"

Use of the term inductorium for the device shown in Figure 2 of the diagram in Tesla's Colorado Springs notes entry for June 27, 1899 (p. 29 of the Interference text) was archaic for this commonplace induction coil even in Tesla's time. An earlier widely used name was Ruhmkorff coil, also attributed to Charles G. Page in the United States. Tesla undoubtedly chose to use a generic term for the iron-core coil to distinguish it from his air-core induction coils. In early automobile days, spark-plug ignition was provided by the familiar Ford® "Model-T" coil. This coil has been used by young boys though the years in providing excitation for small Tesla coils.

A diagram of the iron-core induction coil is shown below. The primary winding is operated from a battery in series with an interrupter. When current flows through the primary, the magnetization in the iron core momentarily opens the interrupter contacts cutting off the magnetization. The process is repeated rapidly. When the current in the primary is interrupted, the sudden collapse of magnetic flux (di/dt) induces an emf in the secondary winding of several thousand volts sufficient to bridge the gap. The condenser provides not only a more abrupt interruption of the primary current, thereby producing greater emf in the secondary, but also serves to reduce arcing at the interrupter contacts which would otherwise suffer erosion.

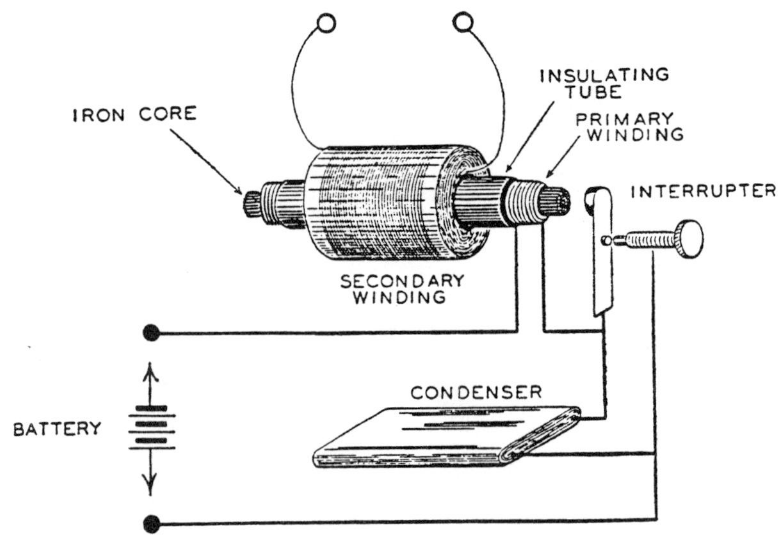

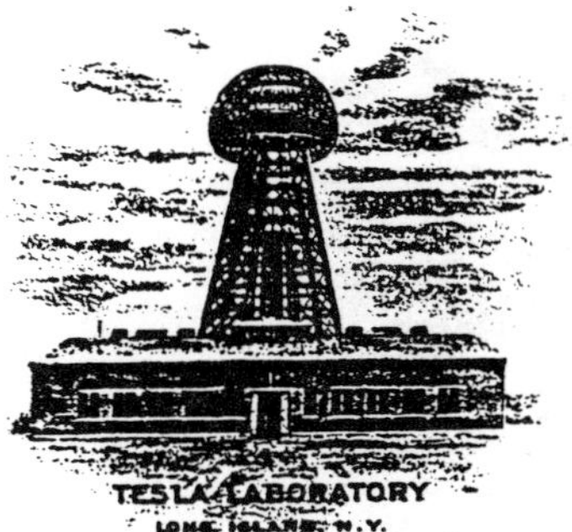

8 West 40th Street
New York, N. Y.
September 29, 1915

Benjamin F. Miessner, Esq.,
Purdue University,
Lafayette, Indiana

My dear Sir:

Your favor of September 24th has been received in due course and has interested me in view of your forthcoming book on "Radio Dynamics". Some time ago my friend, Charles E. Speirs of the D. Van Nostrand Company, told me that you were engaged in its preparation and I commended it for publication as very little has been written on the subject. Personally, I believe that the name is not the very best as it conveys the idea that radiations are, if not motive, at least the controlling agent, while, as a matter of fact, such is not the case.

I am naturally greatly absorbed in this field of invention which has been barely touched and which I look upon as extremely promising. In an article in the Century Magazine, copy of which I am forwarding to you, I have related the circumstances which led me to develop the idea of a self-propelled automaton. My experiments were begun sometime in '92 and from that period, on, until '95, in my Laboratory at 35 South Fifth Avenue, I exhibited a number of contrivances and perfected plans for several complete telautomata. After the destruction of my Laboratory by fire in '95, there was an interruption in these labors which, however, were resumed in '96 in my new Laboratory at 46 East Houston Street where I made more striking demonstrations, in many instances actually transmitting the whole motive energy to the devices instead of simply controlling the same from distance. In '97 I began the construction of a complete automaton in the form of a boat, which is described

TESLA LABORATORY
LONG ISLAND, N.Y.

Benjamin F. Miessner, Esq. -2-

in my original patent specification #613,809. A copy of this, also, is being forwarded under separate cover. This application was written during that year but the filing was delayed until July of the following year, long before which date the machine had been often exhibited to visitors who never ceased to wonder at the performances. The drawings of this specification were made from this machine to scale. In that year I also constructed a larger boat which I exhibited, among other things, in Chicago during a lecture before the Commercial Club. In this lecture I treated the whole field broadly, not limiting myself to mechanisms controlled from distance but to machines possessed of their own intelligence. Since that time I have advanced greatly in the evolution of the invention and think that the time is not distant when I shall show an automaton which, left to itself, will act as though possessed of reason and without any wilful control from the outside. Whatever be the practical possibilities of such an achievement, it will mark the beginning of a new epoch in mechanics.

I would call your attention to the fact that while my specification, above mentioned, shows the automatic mechanism as controlled through a simple tuned circuit, I have used individualized control; that is, one based on the co-operation of several circuits of different periods of vibration, a principle which I had already developed at that time and which was subsequently described in my patents #723,188 and 723,189 of March, 1903. The machine was in this form when I made demonstrations with it in 1898 before the Chief Examiner, Seeley, prior to the grant of my basic patent on Method of and Apparatus for Controlling Mechanisms at a Distance.

Benjamin F. Miessner, Esq. -3-

In my experiments and investigations in Colorado from 1899 to 1900, I developed, among other things, two important discoveries which will be essential in the future development of telautomatics. They are described in my patents #685,953 and 119,732 which were taken out at a later date. These two advances make it possible to supply to an automaton great amounts of energy and also to control it with the utmost accuracy when it is entirely out of sight and at any distance.

During the past few years I have devoted much of my time to the perfection of a small, high speed vessel and have developed a new form of prime mover which makes it possible to develop several horsepower for each pound of weight and in my latest designs I am embodying this new machine together with certain new means of propulsion in an endeavor to produce a most effective weapon of defense, such as would seem to be at this time of paramount importance to the United States.

I may be able to respond to your request to furnish you one or two illustrations but am so driven with important work that it would be next to impossible for me to prepare material, myself, for publication in your book which I hope will prove a complete success.

Yours very truly,

N. Tesla

P/S- I have added to the material forwarded, a few other specifications which might be of interest to you in this connection.

8 West 40th Street
New York, N. Y.
October 8, 1915

Benjamin F. Miessner, Esq.,
Purdue University,
Lafayette, Indiana

Dear Mr. Miessner:

I have duly received your favor of the third instant and wish to thank you for your appreciation of my work, as well as for the delightful frankness with which you have expressed yourself in regard to my mental balance. This is news to me as all my intimate friends, some of whom are men of great achievement, are never ceasing in telling me just the opposite.

It also pleases me to note your receptiveness and enthusiasm which augurs well for your future success.

Perhaps you are right in the choice of the title for your work. My objection to the word is that it implies control by radiations while the art requires a different agency for successful practice.

You are, of course, welcome to make such quotations from my records as you see fit. I will probably be able to send you a couple of photographs giving views of automata I constructed and exhibited.

As to your kind offer to honor me, I am very much obliged to you for the same but I believe that it will be much better for the success of your book not to give too much prominence to any contemporary. Mr. Speirs is a

Benjamin F. Miessner, Esq. -2-

man of unerring judgment in this respect and you can do no better than follow his advice.

Wishing you the best success in what you have undertaken and looking to the pleasure of meeting you, personally, some time, I remain,

Very truly yours,

N. Tesla

8 West 40th Street
New York, N. Y.
November 8, 1915

B. F. Miessner, Esq.,
Purdue University,
Lafayette, Ind.

Dear Mr. Miessner:

In arranging my correspondence in a new office just installed, I ran across a letter from you dated July 2d, 1914 which, so far as I can find, has not been answered and must have been overlooked. Please accept my apologies for the seeming neglect. It is my rule to always answer promptly communications from inventors and members of my profession.

In accordance with promise expressed in my letter to you of October eighth, I have had three prints prepared showing two of my telautomata which have been exhibited on fre-cuent occasions from 1897 to 1899. They have been forwarded to Mr. Sreirs of the D. Van Nostrand Company who will no doubt communicate with you in regard to them.

Believe me,

Yours very truly,

N. Tesla

From the Publisher

The name Nikola Tesla is most often associated with a unique type of radio frequency oscillator known as the Tesla coil, an advanced form of which is shown below in the Wardenclyffe laboratory on eastern Long Island. Hanging from the building's roof braces is a quarter-wavelength spiral coil that was connected to the adjacent helical resonator by the heavily insulated cable.

TESLA WARDENCLYFFE PROJECT ARCHIVES

Development of the Tesla coil began about 1889 as part of the inventor's effort to solve a problem that the scientific community of the time had been wrestling with—how to transmit electrical energy to a distance without wires. In its earliest form the device incorporated a set of series-connected closed core transformers, however, these imposed severe limitations on potential and frequency. The next step was the adoption of an open core transformer and through gradual modifications in the coils' insulation and manner of winding, a much higher level of performance was eventually attained. It was at this point that Tesla eliminated the transformer's iron core which continued to limit the tension which could be developed. The new transformer design had a secondary which was nearly a quarter wave length long, relative to its operating frequency. One of the transformer's terminals was connected to ground with the high potential being produced at the free terminal. A further refinement involved the connection of an additional coil, i.e., a slow wave helical resonator, to the high potential terminal. The 1898 photograph which appears on page 12 of the Interference text is the earliest known appearance of the extra coil magnifier configuration.

Tesla was the last of those great 19th century wireless pioneers who ventured forth to explore the seldom trodden fields of electrical phenomena. In just a matter decades after completing his formal education he had made significant contributions to the early collaborative effort which carried mankind from the horse & buggy, gaslight era into our present age of electricity.

In 1917 Tesla was awarded the Edison Medal by the American Institute of Electrical Engineers, accepting it only at the very last moment. By that time the impractical direct-current system, which was introduced by Thomas Edison in 1882, had been superseded by the vastly superior polyphase alternating current system of Tesla's own design. Guglielmo Marconi received the Nobel Prize for radio in 1909. In 1943, the year of Tesla's death, Marconi's single patent on radio was invalidated by the U.S. Supreme Court because the technology it protected had been anticipated by Tesla.

John Bardeen, Walter Brattain, and William Shockly are properly credited with the invention of the transistor as a replacment for the electronic vacuum tube in analog communications. It is interesting to note that both Tesla, with his art of individualization invention, and Mr. Bardeen, Mr. Brattain, and Mr. Shockly with their transistor invention, found initial applications in communications. While Bardeen, Brattain, and Shockly have received a Nobel Prize for their work leading to the invention of the transistor, Tesla has been denied the highest awards for his discovery of the rotating magnetic field, radio, and the AND logic gate—an invention which is the essential element of all digital computer systems, a leading industry of today. Tesla once poignantly remarked to another one suffering the same fate, "The contemporaries will never understand us—we are too far ahead of our time."

Index